RAND NATIONAL SECURITY RESEARCH DIVISION

T0306388

Developing a Capacity Assessment Framework for Marine Logistics Groups

Joslyn Hemler, Yuna Huh Wong, Walter L. Perry, Austin Lewis

Prepared for the United States Marine Corps

Approved for public release; distribution unlimited

For more information on this publication, visit www.rand.org/t/RR1572

Library of Congress Cataloging-in-Publication Data is available for this publication.
ISBN: 978-0-8330-9668-5

Published by the RAND Corporation, Santa Monica, Calif.

Cover: U.S. Marine Corps photo by Sgt. Paul Peterson.

Preface

In 2006, Force Service Support Groups were redesignated as Marine Logistics Groups (MLGs). The MLGs provide general and direct logistics combat service support to U.S. Marine Corps units. However, many of the resulting structural changes led to uncertainty at both the operational and tactical levels concerning how to measure logistics capacity. The Operations Analysis Directorate (OAD) at Marine Corps Combat Development and Integration Command asked the RAND Corporation to develop a standardized method to determine MLG capacity to provide logistics support. This document reports on the initial development of a standard method.

This research was sponsored by OAD and was conducted jointly within the International Security and Defense Policy and the Acquisition and Technology Policy Centers of the RAND National Defense Research Institute, a federally funded research and development center sponsored by the Office of the Secretary of Defense, the Joint Staff, the Unified Combatant Commands, the Navy, the Marine Corps, the defense agencies, and the defense Intelligence Community.

For more information on the International Security and Defense Policy Center, see www.rand.org/nsrd/ndri/centers/isdp or contact the director. For more information on the Acquisition and Technology Policy Center, see www.rand.org/nsrd/ndri/centers/atp or contact the director (contact information for both directors can be found on their respective web pages).

Contents

Figures and Tables

Figures

Tables

Summary

In 2003, the U.S. Marine Corps restructured its logistics support units in order to provide tailored, functional logistics support to the Ground Combat Element (GCE). The Force Service Support Groups (FSSGs) were redesignated as Marine Logistics Groups (MLGs) in 2006 as part of the Marine Corps' logistics modernization initiative.

Approach

The new MLG structure provides general and direct logistic combat support across all functional areas of logistics. However, MLGs do not have a standardized method to determine their ability to provide logistics support. Each level of command provides a subjective assessment of *logistics capacity* based on experience and knowledge. However, this leads to uneven assessments between units and uncertainty. A standardized methodology and objective analysis would allow the MLGs to assess their logistics capacity in a structured way and enable leaders to communicate their ability to provide logistics support to the supported commander.

Research Objective, Tasks, and Method

The research objective was to provide a capacity assessment framework that would assist each of the MLGs, or any unit of any size, in determining the ability of logistics units to meet current and projected tasks. Research began with the development of logistics units of employment (LUEs) primarily composed of the personnel and equipment required to accomplish a specific task and determine metrics associated with measure output. These LUEs are considered the basic building blocks of logistics capacity and constitute the elements of a framework.

The tasks were to

- develop unit tables of organization and equipment (T/Os and T/Es) for the LUEs
- associate LUEs with mission-essential tasks (METs)
- establish metrics to gauge how well tasks could be supported
- develop a framework for assessing logistics capacity adaptable for any unit
- test the framework for assessing logistics capacity on a Marine Expeditionary Unit (MEU) Combat Logistics Battalion (CLB).

The study proceeded along four steps:

- **Construct the LUEs:** We consulted with the study sponsor and stakeholders to gain perspectives on relevant metrics. We began with the METs, established associated LUEs (see list of the LUEs in Appendix A), and identified metrics for each.
- **Test the LUEs:** We tested the LUEs against CLB-11 and were able to express the unit's logistics capacity in LUEs.
- **Apply the LUE construct to two MEU CLB missions:** We identified the basic LUEs that would represent a beginning point for assessing logistics support to these missions.
- **Extend the LUE framework:** We extended the LUE framework to assess logistics capacity at the MEU CLB mission level as a starting point for translating LUEs to mission-level capacity metrics. Noncombatant evacuation operations (NEOs) and humanitarian assistance and disaster relief (HADR) were selected missions for the study.

Marine Logistics Group

The original structure of the FSSG was to create functional logistics battalions to support the Marine division. This functional structure required constant task organization during deployment as each unit required support from across a variety of different logistics battalions. While the FSSG was able to amass critical logistics components, it was not flexible enough to respond quickly. In addition, the functional structure created no habitual relationship between supported and supporting units.

When the FSSG deployed in support of Operation Iraqi Freedom I, the commanding general decided to task-organize detachments to provide direct support (DS) combat service support (CSS); his intent was to create DS units that were multifunctional, rapidly deployable, and that maintained habitual relationships with supported units. The reorganization proved successful, and the Marine Corps in 2006 implemented a series of logistics modernization initiatives, one of which was a restructuring of the FSSG to the MLG.

MLG Organization and Operations

The current MLG structure supports both direct and general support (GS) logistics capabilities. It is built around standing Combat Logistics Regiments (CLRs) and CLBs. Within 1st MLG, there is a headquarters regiment, a DS CLR, a GS CLR, an engineer support battalion (ESB), a medical battalion, and a dental battalion.

- **Headquarters regiment:** The headquarters regiment is responsible for most of the services functions of the MLG and contains CLBs that support MEUs. It is responsible for providing command and control (C2) functions required to integrate services support.
- **CLR-1:** CLR-1 is the DS regiment for 1st MLG. Its mission is to provide logistics support beyond 1st Marine Division's organic capability.
- **CLR-15:** GS logistics functions are conducted by CLR-15. Its mission is to provide intermediate supply, field-level maintenance,

materiel distribution support, procurement management, equipment fielding support, and forward resuscitative health care to the Marine Expeditionary Force (MEF).
- **7th ESB, 1st Medical Battalion, and 1st Dental Battalion:** 7th ESB provides general engineering support to the MEF. 1st Medical Battalion provides health service support and 1st Dental Battalion is the source of dental services to the MEF.

In addition to the fixed MLG structure, the MLG also provides logistics combat elements (LCEs) to smaller Marine Air Ground Task Forces (MAGTFs). One of the principle operating concepts of the Marine Corps in general is to generate MAGTFs that are task-organized, rapidly deployable, and self-sustainable.

The MLG is also responsible for supporting both its own internal logistics readiness as well as external service logistics readiness requirements.[1] The MLG conducts 47 METs. It is tasked to perform all six functions of logistics: transportation, general engineering, supply, maintenance, health services, and services.

Impact of New Organization on Logistics Support

Under the FSSG construct, the Marine Corps employed a functionally aligned logistics structure that was capable of providing economies of scale to support the MEF. While the FSSG was able to support intermediate-level logistics requirements levied by the MEF, it lacked multifunctional units with habitual relationships between supported and supporting units.

The MLG is closely aligned with key Marine Corps strategic concepts that call for smaller-scale forces that can expand rapidly. While the new structure supports these concepts, the change in organization

---

[1] Readiness, as discussed in this document, follows the definition in Joint Staff, U.S. Department of Defense, Joint Publication 1-02, *Department of Defense Dictionary of Military and Associated Terms*, Washington, D.C., November 2010: "The ability of military forces to fight and meet the demands of assigned missions." It does not refer to status within the Defense Readiness Reporting System.

has left many logisticians and planners confused as to the ability of the MLG to meet its operational logistics demands. For example, what is the transportation capacity of a DS CLB? How many missions can be supported under the current structure? These are important questions whose answers help planners properly allocate assets.

Determining Logistics Capacity Metrics

For the purpose of this report, *logistics capacity* is defined as the measurement or estimate of the amount of output produced by a logistics system per unit time. Therefore, an LUE is a system whose output is measured in terms of a single measure. By measuring output, *capacity* is defined as the output of a single unit as determined by the manpower and equipment utilized.

In order to develop metrics useful to the logistics community, we interviewed stakeholders identified by the sponsor, primarily the Study Advisory Committee. These discussions were aimed at better understanding the types of logistics capacity metrics that would be the most useful to different stakeholders. We were able to hear from Installations and Logistics (I&L), 2nd MLG, and Marine Forces Command (MARFORCOM), in addition to 1st MLG. In general, stakeholders have different levels of logistics capacity that they are most concerned about.

Higher organizations within the Marine Corps, such as I&L and MARFORCOM, require more aggregated capacity metrics. They are interested in quantifying logistics capacity to address such issues as readiness and force management. MARFORCOM wanted to understand the "residual capacity" of the MLGs to support additional missions after accounting for units that were already deployed. Similarly, I&L was interested in identifying the logistics force structure that was committed, and then understanding the capacity of what was left.

However, the MLGs expressed a desire to better communicate logistics capacity to GCE commanders. Logistics metrics in this case are higher fidelity than those needed for Marine Corps–level force management: They focus on direct connections to specific tasks. The MLGs also discussed the importance of having logistics capacity metrics that were tied to the METs.

Ideally, metrics should be constructed to support different command levels. At lower levels, LUEs can serve to approximate standardized logistics packages and aid in discussions between LCE and GCE commanders. Either further aggregation of the LUEs or a top-down heuristic separate from the LUEs is needed to create metrics that are useful to higher echelons.

Levels of Metrics for Logistics Capacity

There are different levels of logistics-capacity measurements. The levels of metrics are envisioned as a pyramid, with more aggregate measures relying on the levels below them:

- The *planning factor* is at the bottom of the pyramid. Units of measurement are the number of passengers accommodated, the gallons produced, and more.
- At the *mission-planning* level, metrics are aggregated and mapped to identifiable tasks to produce the LUEs. Taking from the earlier examples, the metrics for this level would still be passengers accommodated or gallons produced, but it would be the total based on the amount of equipment and personnel in the LUE.
- At the top, or *supportable-missions* level, the focus is on force management and readiness. LUEs are aggregated by identifying different combinations of LUEs with each mission type. The objective is to identify the total number of missions (such as HADR or NEO) that can be supported.

Associating Mission-Essential Tasks to Logistics Units of Employment

Measuring the logistics capacity of a unit requires two aspects: (1) well-defined terms and output metrics; and (2) resources to ensure the unit is capable of effective utilization of its capacity. The LUE is the building block needed to measure a logistics task. It can be expanded to provide greater logistics capacity. Each LUE is associated with a metric that measures its capacity to perform its associated task. In order to create LUEs, it is important to understand the core logistics capabilities of the MLG.

Methodology

LUEs were created to support 1st MLG's mission-essential task list. Once all key tasks were identified, a review of Marine Corps warfighting doctrine was conducted to identify those building block units that already exist. LUEs were then validated with subject-matter experts (SMEs). The end result was a list of LUEs that cover all METs required of any of the four MLGs.

Better understanding logistics capacity consists of incorporating information on three key components: personnel, equipment, and appropriate metrics. The next step—once the three logistics capacity components were assigned to each LUE—was to review unit T/Os and T/Es. This determined the number of LUEs that could be fielded by each unit. Complicating this process is the fact that many pieces of equipment are dual use or alternate pieces of equipment can be used in lieu of the primary. Because of time and resource constraints, this validation was conducted for a single unit, CLB-11, the logistics battalion within 1st MLG that supports 11th MEU.

Limitations

To perform certain tasks, personnel within an LUE should be of a certain rank, possess certain training, or acquire certain licenses, i.e., the unit must be *capable* of effectively utilizing the capacity available. Capability details are not included in the LUE. The Marine Corps tries to fill billets with personnel who fit these requirements, but sometimes it is forced to fill with lesser ranks or a unit has trouble sending the Marine to training.

LUEs also contain limited ability to demonstrate the substitutability of personnel and equipment. The LUEs for the most part contain the optimal piece of equipment preferred by doctrine and by SMEs. However, the LUEs do provide capacity information for alternate pieces of equipment that reside within the MLG.

Functions of Logistics

Table S.1 includes a summary description of the six logistics functions.

Table S.1
Functions of Logistics

Logistics Function	Description
Transportation	Transportation operations distribute supplies across various means. The key subfunctions are motor transport, materiel handling, landing support, embarkation, freight and passenger transportation, aerial delivery, and port and terminal operations.
General engineering	Engineering personnel and assets within the ESB provide expeditionary vertical and horizontal construction, standard and nonstandard bridging, explosive ordnance disposal support, bulk fuel operations, bulk water operations, and tactical utility support.
Supply	Organic support includes limited administrative functions: fiscal, due and status file management, consolidated memorandum receipt management, and basic warehousing. Intermediate support consists of repairable management, supply chain materiel management, ground supply operations, requirements determination, procurement, storage, and distribution operations.
Maintenance	Maintenance is divided into two sublevels: field and depot. The field level includes organizational and intermediate maintenance, and is the responsibility of the using unit. Intermediate maintenance requires a higher level of technical capability than the organizational level; and depot maintenance requires greater capability than the field level of maintenance.[a]
Health services	Health-services tasks consist of casualty treatment, temporary casualty holding, casualty evacuation, dental health readiness, emergency dental services, and medical regulating.
Services	Services consist of both CSS and combat service support element (CSSE) services. CSS are those services inherent to a command, such as personnel administration, religious ministry, billeting, financial management, morale, welfare and recreation, and messing. CSSE services are those not available or organic to other MAGTF elements, including postal, legal, field exchange, disbursing, mortuary affairs, and civil-military operations.[b]

Table S.1—Continued

[a] Organizational tasks consist of inspecting, servicing, lubricating, adjusting, replacing of parts, minor assemblies, and subassemblies. Intermediate maintenance requires a higher level of technical proficiency. Tasks here include modification, replacement, fabrication, component/subcomponent/assembly/subassembly repair, calibration and repair of test, measurement and diagnostic equipment, software maintenance, precision machining, welding, evacuation, disposal, salvage, and demilitarization of equipment or materiel. Depot-level tasks include inspection, repair, overhaul, or the modification or reclamation of weapons systems, equipment end items, parts, components, assemblies, and subassemblies beyond field maintenance capabilities. For further information, see Headquarters, U.S. Marine Corps, *Ground Equipment Maintenance Program*, Marine Corps Order 4790.25, Washington, D.C., January 12, 2014a.

[b] Operational Contract Support (OCS) is an important service that is considered a separate logistics function under Joint doctrine, but it was outside the scope of this study. For further information on OCS, please see Joint Staff, U.S. Department of Defense, Joint Publication 4-10, *Operational Contract Support*, Washington, D.C., July 16, 2014a.

Logistics Units of Employment Framework

The term *framework* here means a basic structure underlying a system or concept. It is intended to support or guide the extension of the basic structure into useful constructs. In this respect, the LUEs themselves represent a basic, extendable framework.

Two applications of the LUEs as a framework are illustrated. The first is in using them to approximate the capacity of a Marine Corps logistics organization. The second application of the LUEs as a framework is in using the LUEs to understand the logistics capacity required for MEU CLB support to HADR.

Measuring the Capacity of CLB-11

LUEs offer a way to standardize logistics capacity measures. Comparing LUEs that different units should be able to support provides a first-order approximation of logistics capacities across organizations. It is also possible to extend this to each MLG's entire structure: Adding up all of the LUEs that its total structure can support will produce an approximate total capacity in each logistics functional area. It is then possible to see how many LUEs are dedicated to deployed forces, overall MEF support, and more at a given point in time.

We chose CLB-11 from 1st MLG to demonstrate the use of the LUEs as capacity metrics. The process involved examining the T/O and T/E for CLB-11 and assessing the number of LUEs that it could support.

Measuring Logistics Capacity for an MEU Mission

We can also use the LUEs to estimate the likelihood that a given logistics organizational structure is sufficient to support a mission. We do this by comparing the types of LUEs needed for a mission with the unit's available LUEs.

Conclusions and Recommendations

Conclusions focus on the development and application of the LUEs for understanding logistics capacity whereas recommendations focus on further improvement of the LUEs and areas for potential further development.

Conclusions

Based on the results of this work, we found that it is feasible to create a framework for understanding logistics capacity for the Marine Corps. The LUEs represent the basic building blocks of logistics capacity—the smallest set of personnel and equipment needed to accomplish essential tasks. The LUEs can be applied across each of the functional areas of Marine Corps logistics, and can be used as a basic, extendable framework for understanding logistics capacity across the MLG. Because a given logistic unit's T/O and T/E can be expressed as LUEs, the LUEs provide a means to gauge and compare the capacity of different units.

This work also demonstrates the hierarchy of frameworks that can be applied for different purposes and at different organizational levels. In addition, it is possible to identify the LUEs associated with different missions as an extension of the basic LUE framework. Identifying the LUEs needed for a mission with the LUEs available from within an organization provides information on the number and type of missions the organization might support. The extension of the basic LUE framework to HADR and NEO missions illustrates this point.

Recommendations

This study represents a promising start to developing a capacity assessment framework for MLGs. LUEs were created for the MLG and the framework was applied to a single unit (CLB-11) and two missions (HADR and NEO); however, much more remains to be accomplished before the LUEs listed in Appendix A can be validated. In the future, we suggest that additional work be conducted in these areas:

- **Expand the LUE framework:** The current version of the LUEs does not take *grade* or *training* requirements into consideration when accounting for personnel. They also have limited information about substitutes for the equipment listed. Adding these components would greatly improve their assessment capability.
- **SME LUE input:** A significant limitation of the LUEs is the nature of the SME input used to create many of them. In cases where there was little written guidance on the necessary personnel and equipment to accomplish a task, we relied heavily on SME judgment. Moreover, the SMEs used in this study were limited to only a very narrow sample of Marine Corps logistics experts; broader sampling of experts and examining the actual employment of personnel and equipment are needed.
- **Expand the application of the LUE framework:** Apply the basic LUE framework to the entire T/O and T/E of 1st MLG or another MLG. A logical next step is to turn the LUEs into a logistics estimation tool or dashboard that would demonstrate capacity based on current unit resources, which the Marine Corps is currently developing. Gathering information on the number of LUEs needed to support the MEF outside of deployments would lead to better understanding of residual capacity.
- **Expand the mission set:** Identify the approximate number of LUEs needed for each CLB mission. This would provide valuable information in assessing readiness and force management

and would better assess the number of missions an MLG might potentially support.

- **Quantify capacity:** Develop a better method to quantify capacity in harder-to-measure areas, such as the supply and maintenance functional areas. The GS requirements for supply and maintenance proved particularly difficult in capturing capacity and determining building block–level requirements similar to other LUEs.
- **Logistics C2:** C2 was not dealt with in this study. Potential limitations stemming from C2 issues are not currently accounted for in the LUEs, but they would be an important issue to examine in greater detail.
- **Self-sustainment:** Account for the logistics support that logistics units require to sustain themselves, even as they provide support to other MEUs and entities outside the Marine Corps. Because this is not currently captured, the true logistics capacity required for the Marine Corps might be systematically underestimated.

Acknowledgments

The authors wish to acknowledge the considerable assistance provided by Maj. Joseph Maddux, Robert Hayden, Brittlea Brown, and Al Sawyers at Marine Corps Operations Analysis Directorate. The study also benefited immeasurably from the assistance provided by CWO3 Andre Mayes at 1st Marine Logistics Group. We are also grateful to the other members of the Study Advisory Committee and other stakeholders for their input and guidance, including Robert Lain at Marine Forces Command, LtCol. Dean Stouffer at 2nd MLG, LtCol. Omar Randall at Installations and Logistics (I&L), LtCol. Daniel Reber at I&L, Michael Resnick at I&L, Frank Rosa at I&L, and LtCol. Brian Clemens at Marine Corps Forces, Pacific.

CHAPTER ONE

Introduction

In 2003, as the 1st Marine Expeditionary Force (MEF) was preparing for its deployment in support of Operation Iraqi Freedom (OIF), the U.S. Marine Corps decided to restructure its logistics support units in order to provide tailored, functional logistics support to the Ground Combat Element (GCE). The Force Service Support Groups (FSSGs) were permanently reorganized and redesignated as Marine Logistics Groups (MLGs) in 2006 as part of Marine Corps logistics modernization initiative. The new MLG task organization provides general and direct logistics combat support across all functional areas of logistics. However, many of the resulting structural changes, as well as the loss of 2,300 billet identifier codes, resulted in uncertainty at both the operational and tactical levels in providing informed decisions about logistics capacity.

Currently, MLGs do not have a standardized method to determine their ability to provide logistics support. Each level of command provides a subjective assessment of logistics capacity based on experience and knowledge. However, this subjective approach leads to uneven assessments and uncertainty at the operational and tactical levels.

Operational planners must clearly articulate the ability of a unit to meet task and mission objectives. However, the subjectivity of logistics capacity is merely one challenge facing logisticians eager to provide informed decisions. Other challenges compounding the assessment include the variation in size, composition, and organization of each MLG. In addition, the increased demand for smaller detachment-sized units to support enduring forward presence, global and crisis response, and distributed operations highlights the requirement for articulating capacity below the battalion level. The disaggregated operations called for in *Expeditionary Force 21* will also further stress Marine Corps logistics capacity when it comes to high-demand and low-density capabilities, such as power distribution, transportation, and materiel handling.[1]

A standardized methodology and objective analysis would allow the MLGs to assess their logistics capacity in a structured way, enabling leaders to effectively communicate their ability to provide logistics support, particularly to the supported commander.

Research Objective, Tasks, and Method

The project's overall research objective was to provide a capacity assessment framework for all four MLGs to assist in determining the ability of logistics units to meet current and projected tasking. Because the question of Marine logistics capacity involves multiple layers of metrics and data collection, we began with creating logistics units of employment (LUEs). The term *LUE* was conceived by the sponsor during the course of this project, so it is not found in Marine Corps doctrine. For the purpose of this report, an LUE is primarily composed of the personnel and equipment required to accomplish a specific task and the metrics associated to measure output. These LUEs are considered the basic building blocks of logistics capacity and constitute the elements of a framework. This report also proposes a way ahead for further development of logistics capacity metrics.

The LUEs offer a way to standardize logistics capacity measures across different logistics organizations. Comparing the LUEs that different units should be able to support provides a first order approximation of logistics capacities across organizational structures. It is also possible to extend this to an MLG's entire structure: Adding up all the LUEs that its total structure can support will produce an approximate

[1] Headquarters, U.S. Marine Corps, *Expeditionary Force 21*, Washington, D.C., March 2014b.

total capacity in each logistics functional area. It is then possible to see how many LUEs are dedicated to deployed forces, overall MEF support, or are resident in residual logistics capacity at a given point in time.

The tasks initially identified by the sponsor, Marine Corps Operations Analysis Directorate (OAD), were to (1) develop unit tables of organization and equipment (T/Os and T/Es) for the LUEs; (2) associate LUEs with mission-essential tasks (METs); (3) establish metrics to gauge how well tasks could be supported logistically across a range of Marine Air Ground Task Force (MAGTF) operations; (4) develop a framework for assessing logistics capacity adaptable for any unit; and (5) test the framework for assessing logistics capacity on a Marine Expeditionary Unit (MEU) Combat Logistics Battalion (CLB), time permitting. Logistics within the aviation combat element were excluded from this report.

We consulted with the study sponsor and stakeholders to gain perspectives on the types of metrics that would be most relevant. We began with the METs, developed associated LUEs for the METs, and identified appropriate metrics for each of the LUEs. We also tested the LUEs against CLB-11 and were able to express the unit's logistics capacity in LUEs. The team also applied the LUE construct to two MEU CLB missions, and identified the basic LUEs that would represent a beginning point for assessing logistics support to these missions. The RAND Corporation Human Subjects Protection Committee determined this work not human subjects research, and the U.S. Marine Corps Institutional Review Board concurred.

Constructing Logistics Units of Employment

We developed initial LUEs by reviewing existing documentation on logistics, and then consulted with subject-matter experts (SMEs) with backgrounds in the six subfunctions of Marine Corps logistics. Revised LUEs were provided to OAD and the study advisory committee (SAC) for comments, and then provided to the sponsor in the form of an Excel spreadsheet. A description of the LUE spreadsheet is in Appendix A of this report. The LUEs are meant to be applicable across the MLG for logistics in general, and are not limited to the MEU CLB.

We created initial LUEs by reviewing the mission-essential task list (METL) for 1st MLG for each task to create a database of LUEs. A subsequent review of Marine Corps doctrine, warfighting, and reference publications developed further standard units of employment. Relevant publications included engineering, health services support, transportation, maintenance, petroleum and water operations, supply, services, and other logistics publications.[2]

After developing this initial set of LUEs, we conducted a week-long field visit to 1st MLG at Camp Pendleton in January 2016. As one of the stakeholders who had initiated the study, 1st MLG hosted the visit and identified logistics experts. 1st MLG arranged for small-group discussions and individual interviews with the identified experts in each of the subfunctional areas. Subfunctional area discussions lent additional insights into the identified tasks and assisted in the further development of T/Os and T/Es for each LUE. We were also able to conduct follow-on phone conversations with a number of the SMEs after the visit on additional questions that came up as the team continued to work on the LUEs.

Another step in the process was to establish metrics for logistical support. Many of the METs came with an associated "metric," but some

---

[2] Headquarters, U.S. Marine Corps, *Engineering Operations,* Marine Corps Warfighting Publication (MCWP) 3-17, Washington, D.C., February 14, 2000a; Headquarters, U.S. Marine Corps, *Logistics Operations,* MCWP 4-1, Washington, D.C., April 15, 1999; Headquarters, U.S. Marine Corps, *Tactical-Level Logistics,* MCWP 4-11, Washington, D.C., June 13, 2000b; Headquarters, U.S. Marine Corps, *Health Service Support Operations,* MCWP 4-11.1, Washington, D.C., December 10, 2012b; Headquarters, U.S. Marine Corps, *Transportation Operations,* MCWP 4-11.3, Washington, D.C., September 5, 2001a; Headquarters, U.S. Marine Corps, *Maintenance Operations,* MCWP 4-11.4, Washington, D.C., April 24, 1998a; Headquarters, U.S. Marine Corps, *Bulk Liquid Operations,* MCWP 4-11.6, Washington, D.C., June 19, 2005; Headquarters, U.S. Marine Corps, *MAGTF Supply Operations,* MCWP 4-11.7, Washington, D.C., June 24, 1998b; Headquarters, U.S. Marine Corps, *Services in an Expeditionary Environment,* MCWP 4-11.8, Washington, D.C., September 24, 2001b; Headquarters, U.S. Marine Corps, *Logistics,* Marine Corps Doctrinal Publication 4, Washington, D.C., February 21, 1997; Headquarters, U.S. Marine Corps, *Unit Embarkation Handbook,* Marine Corps Reference Publication 4-11.3G, Washington, D.C., December 10, 2004; Headquarters, U.S. Marine Corps, *Marine Corps Field Feeding Program,* Marine Corps Reference Publication 4-11.8A, Washington, D.C., December 2, 2013; Headquarters, U.S. Marine Corps, *Organization of Marine Corps Forces,* Marine Corps Reference Publication 5-12D, Washington, D.C., August 26, 2015.

were such instructions as "establish within 48 hours of arrival" rather than capacity-related metrics.[3] To meet the study objective of establishing metrics, as requested by the sponsor, we used analyst judgment and SME input to assign "units of measure" for the LUEs. The intent of these units of measure is to identify capacity metrics that logisticians could use to communicate capacity to the GCE commander. Example units of measure include the number of passengers or tons that can be moved, number of casualties that can be accommodated, or number of requisitions that can be handled.

Extending the Framework for Logistics Capacity for MEU Missions

We then extended the LUE framework to understand logistics capacity at the MEU mission level. Although the LUEs were meant to be broadly applicable across the MLG, the frameworks were focused on the MEU mission level. The focus was on MEU CLB as a starting point for translating capacity building blocks to mission-level capacity metrics. OAD and SAC members identified noncombatant evacuation operations (NEOs) and humanitarian assistance and disaster relief (HADR) from among the available MEU CLB missions of interest. Although there was interest in a third, stability operations (StabOps), the variation in size and diversity of tasks for StabOps presented a challenge.

The RAND team began with the current METL for 1st MLG and identified the tasks relevant for NEO and HADR and identified the logistics-related standards associated with those two missions.[4] The study team then referred to the training and readiness (T&R) manual and identified the tasks required to achieve those standards.[5] The appropriate LUE was then aligned to the T&R tasks. The associated

---

[3] 1st MLG transportation METs (see U.S. Marine Corps, Capabilities Development Directorate, *Marine Corps Task List 2.0*, Quantico, Va., September 1, 2016).

[4] For HADR, these were Marine Corps task (MCT) 1.13.2.1 Provide Evacuation Control Center (ECC), and 1.15.1.2 Facilitate Foreign Humanitarian Assistance. For NEO, the relevant MCT was 1.13.2.1 Provide ECC.

[5] Headquarters, U.S. Marine Corps, *Marine Expeditionary Unit Training and Readiness Manual*, Navy and Marine Corps 3500.99, Washington, D.C., November 13, 2012a.

LUEs for HADR provide the basic building block set for the mission and constitute the logistics capacity framework for this mission. This basic LUE set identifies the minimum number of necessary unique personnel and equipment for HADR, as identified through the MCTs and T&R tasks, and is listed in Appendix C. We also examined the T&R standards related to NEO, but found that there was only one task at the logistics combat element (LCE) level listed, and no LUE that was an immediately obvious fit for the task. Additional discussion is found in Chapter Five.

In order to use the framework to provide a measure of logistics capacity, some idea of the size of the operation and the level of demand for each of the tasks would be necessary to identify the number of LUEs required. To increase the rigor in estimating the quantity and distribution of LUEs, it would be ideal to triangulate between historical data on logistics personnel and equipment used in past operations, demand data for logistics forces, and SME estimates on what should be provided for a range of example operations.

In order to supplement the information from the MCTs and the T&R manual, RAND also held structured brainstorming sessions with SMEs to identify MEU CLB logistics tasks for both MEU missions during the January 2016 visit to 1st MLG. Participants were asked to identify logistics tasks in each subfunction that provided internal support to the MEU CLB, versus logistics support that was externally provided to the MEU, civilians, and others outside the CLB. Participants also arranged the tasks in rough chronological order, going from premovement (covering planning, embarkation, and rehearsal), to movement, to action.

This exercise raised two questions about attempting to measure MEU CLB logistics capacity. The first was the issue of understanding logistics demand over time; meaning any scoping of the capacity frameworks would need to take into account how demand for the same types of personnel and equipment changes over time. In other words, are several of a given LUE typically in demand all at once, or is the demand spread out over the course of the operation? Demand spread out over time means that a smaller number of LUEs are able to provide the same amount of capacity as a larger number of LUEs that must be employed simultaneously.

The second question raised by this brainstorming exercise was the volume of tasks that participants identified as internal logistics support to the MEU CLB itself. Although the exercise was not designed to provide a quantitative ratio of self-sustaining logistics support versus support provided to the MEU and other entities, the question of what additional capacity should be accounted for in order for the MEU CLB to sustain itself is an important one.

Challenges

Several challenges were present for this project. Logistics capacity is a broad topic with multiple stakeholders, and the project's scope only permitted a beginning look at the many issues involved. The lack of centralized and consistent data on requests for Marine Corps capabilities and the deployments of those capabilities was a significant challenge for the project, and will remain a significant one for further work on measuring logistics capacity. Command and control (C2) functions and simultaneity of resourcing in logistics were also challenging issues to incorporate into measuring logistics capacity.

Two areas of logistics proved to be especially difficult to understand from a capacity perspective: supply and maintenance. For supply, there is the distinction between *direct support* (DS) and *general support* (GS). While the capacity needed for DS to a specific unit might be easier to capture, it is more difficult to estimate how the GS capabilities that the MLG needs to support a variety of tasks and units should be measured.

For maintenance, the *levels of maintenance* (LOMs) offer a spectrum of capability to service or even rebuild equipment.[6] At higher echelons, there is an increase in specialized capability to conduct maintenance, as well as a wider number of units that are supported at each echelon. For instance, intermediate maintenance requires a higher level

---

[6] Headquarters, U.S. Marine Corps, 1998a, p. 3-5.

of technical capability than organizational capability. This organizational structure for maintenance complicates efforts to approximate a standard T/O and T/E for maintenance at a mission level. Data on how often units send equipment to higher LOMs would assist in quantifying how often units access different echelons of maintenance. However, some minimum critical mass of specialized skill and equipment are likely required at each more-advanced level. Therefore, required capacity may be nonlinear with the number of units or missions supported.

Lastly, another challenge to understanding logistics capacity through standardized metrics or packages is the "science versus art" of logistics. The variation in individual skill or experience, the situational factors that determine desired capabilities, and the sheer number of environmental factors that affect the performance of Marines and equipment make it difficult to agree on standards. The task-organization ethos of the Marine Corps and the MLG construct of organizing logisticians by functional area further tilt the preference away from identifying standardized logistical packages.

Limitations for the LUEs

Although the LUEs represent useful building blocks for logistics capacity, there are a number of methodological limitations in the way we have approached their creation. The study sponsor specifically expressed the desire to get away from "subjective" judgment in their creation, and to arrive at more objective and analytically rigorous metrics. Additionally, 1st MLG requested that the study not merely re-create what was already available in manuals and other documentation. However, to meet the time constraints with the available data, the LUEs were predominantly built using SME judgment and existing documentation. There were limitations to both the SME expertise and documentation used in this study:

- **SME input limitation:** The SME input used to develop the LUEs is from the relatively small number of experts in each area for any specific topic, such as bridging or staffing for critical care nursing.

It is not known how representative these SME perspectives are within their respective areas. While the team did interview representatives of the SAC from 1st and 2nd MLG, the focus groups were only conducted with 1st MLG SMEs. This meant that the study team could not compare how SMEs at 2nd or 3rd MLG may differ in their recommendations on structuring LUEs. More-systematic sampling of logistics experts throughout the Marine Corps would improve on the methodology used in this study, but requires additional resources.
- **Document limitation:** Much of the Marine Corps doctrine on logistics is old. For example, doctrine on MAGTF supply operations is 20 years old.[7] Available doctrine also often referenced the FSSG,[8] an older organizing concept for Marine Corps logistics used before the current MLG construct.

A more comprehensive approach would also review data on personnel and equipment that the MLGs have used to support the MEFs. An assessment of actual usage of logistics personnel and equipment, in support of deployed forces, as well as those remaining at the MEFs, would be an important data source to compare with SME interviews and documented guidance. Unfortunately, such usage data were not available at a detailed level to contribute to the creation of LUEs.

One final limitation of the LUEs is that they incorporated limited information about the substitutability of personnel and equipment. In other words, what other types of personnel or equipment are close substitutes for the ones listed in the LUE, in case the preferred equipment and personnel are not available? Although the RAND study team attempted to draw out potential substitutes during discussions with logistics SMEs, SMEs typically identified the preferred personnel and equipment.

---

[7] Headquarters, U.S. Marine Corps, 1998b.

[8] Headquarters, U.S. Marine Corps, 2000a, pp. 1–8; Headquarters, U.S. Marine Corps, 1998b, p. 5-2; and Headquarters, U.S. Marine Corps, 1998a, p. 1-3.

Limitations for the MEU CLB Capacity Frameworks

The MEU CLB capacity frameworks also come with methodological limitations. Because the frameworks used the LUEs, they are also subject to the same limitations. Additionally, as mentioned earlier in the MEU CLB method section, the frameworks were based on the MCTs. Therefore, they did not cover any additional logistics capacity needed for MEU CLBs to support themselves.

The frameworks do not get to standard packages of logistics T/Os and T/Es for NEO and HADR. They offer a starting point of capacities needed, but not the quantities of each needed for a standardized package. Additional data are necessary before this could be attempted. However, data collection at either the 1st or 2nd MLG was sporadic. The study team reviewed manning documents from previous MEU and Special Purpose MAGTF (SPMAGTF) deployments, but these documents did not have the logistics requirements for the missions. Without historical logistics requirements, it was not possible to judge whether any available T/Os and T/Es had been adequate to meet the logistics demand signal.

C2 requirements were also not addressed in the MEU CLB frameworks. Discussions with SAC members, particularly conversations with 2nd MLG, raised the issue of providing C2 elements for logistics forces.[9] Limitations on the MLG's ability to provide appropriate C2 are an important capacity constraint to consider. The T&R manual for NEO and HADR also specified C2 tasks for those missions. However, the scope of this project excluded C2.

About This Report

The remainder of this report is divided into five chapters. Chapter Two provides background on the evolution of the MLG, its organization, and its operations. It is important to understand the structure of the MLG and its METs in order to build LUEs or frameworks that reflect

---

[9] Telephone call with a representative from 2nd MLG, September 29, 2015.

logistics capacity. Chapter Three covers the issue of determining appropriate metrics for logistic capacity. It focuses on the differing needs of stakeholders in the logistics community, and the appropriateness of different metrics to meet their needs. This chapter includes frameworks and measures of capacity that other military services, private companies, and other logistics systems utilize. An overview of the input-output model that was used to develop the LUEs is also presented in this chapter. The METs that the LUEs support are presented in Chapter Four. This includes the methodology used, as well as considerations for each of the six functions of logistics. The LUE framework assessment is presented in Chapter Five. We demonstrate use of the LUE framework to gauge logistics capacity by applying it to a CLB, and also use the LUEs to map out basic logistics requirements for MEU CLB missions. Chapter Six is a summary of the project's conclusions and suggested way forward. In addition, there are three appendixes included in this report.

CHAPTER TWO

Marine Logistics Group

The mission of the MLG is "to provide direct support to the MEF GCE and GS and sustained tactical-level logistics support above the organic capabilities of supported elements of the MEF."[1] The Marine Corps has experimented with different task organizations for logistics over the past 30 years to determine the best method of supporting the MEF. This continual reorganization has left many supported commanders and logistics planners confused about the capacity of the MLG. This chapter provides a brief overview of the evolution of the LCE from the functional structure of the FSSG to the more multifunctional structure of the MLG. Understanding the organizational structure and the changes it has undergone is important in determining the unit's capacity.

Transition from FSSG to MLG

The original structure of the FSSG (Figure 2.1) consisted of functional logistics battalions to support the MEF. Therefore, the task organization included a Brigade Service Support Group (BSSG), maintenance battalion, supply battalion, engineer support battalion (ESB), transportation support battalion (TSB), dental battalion, and medical battalion. This functional structure required constant task organization during deployment in Operation Desert Storm to meet operational

[1] 1st Marine Logistics Group, homepage, October 2015.

Figure 2.1
Pre-OIF 1st FSSG Organization

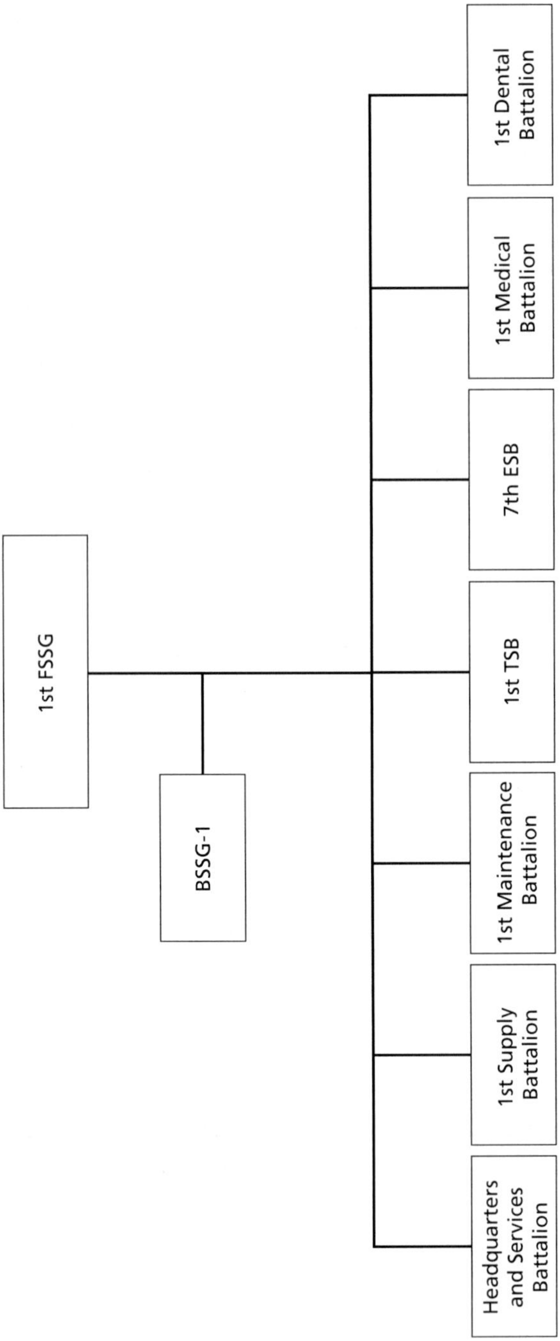

SOURCE: Isabel Marin, Christine Hannigan, Megan Misencik, and Clinton Jones, "Functionally Aligned Battalions," *Marine Corps Gazette*, Vol. 98, Issue 10, October 2014.

RAND *RR1572-2.1*

demands, as each supported unit required support from a variety of different logistics battalions. This involved assessing the needs separately for each unit that required support, and task-organizing logistics personnel and equipment from many battalions to provide that support. A criticism of the structure was that while the FSSG was able to mass critical logistics components, it was not flexible enough to readily respond. In addition, the functional structure created no habitual relationship between supported and supporting units. Advantages of habitual relationships between units include greater communication and established working relationships, while the lack of habitual relationships requires individuals to spend time establishing communication and working relationships.

When the FSSG deployed in support of OIF I (Figure 2.2), the commanding general, Brigadier General Edward Usher, determined that the functional structure of the FSSG, which was successful in garrison, was not suited for supporting the 1st MEF scheme of maneuver. Therefore, he decided to task-organize detachments to provide DS combat service support (CSS), yet still retain a GS capability to support the MEF.[2] This new organization included Combat Service Support Groups (CSSGs), a transportation support group (TSG), and a military police (MP) battalion. His intent was to create DS units that were multifunctional, rapidly deployable, and maintained habitual relationships with supported units. The reorganization proved successful, and the Marine Corps in 2006 implemented a series of logistics modernization initiatives, one of which was a restructuring of the FSSG and a redesignation to the MLG. This new structure closely resembled the force structure implemented during OIF. This was part of a broad move within the Marine Corps to move toward a more modular force.

The new MLG reorganization has had both positive and negative effects on its ability to accomplish its mission. While the MLG is more capable of being rapidly deployable and flexible in employment, a key concept in the post-OIF and Operation Enduring Freedom (OEF)

---

[2] Edward G. Usher, John Sweeney, Darell Moore, Frank Tapia, et al., "Brute Force Combat Service Support: 1st Force Service Support Group in Operation Iraqi Freedom," *Marine Corps Gazette*, Vol. 87, Issue 8, August 2003, pp. 34–35.

Figure 2.2
1st FSSG Task-Organized for OIF I

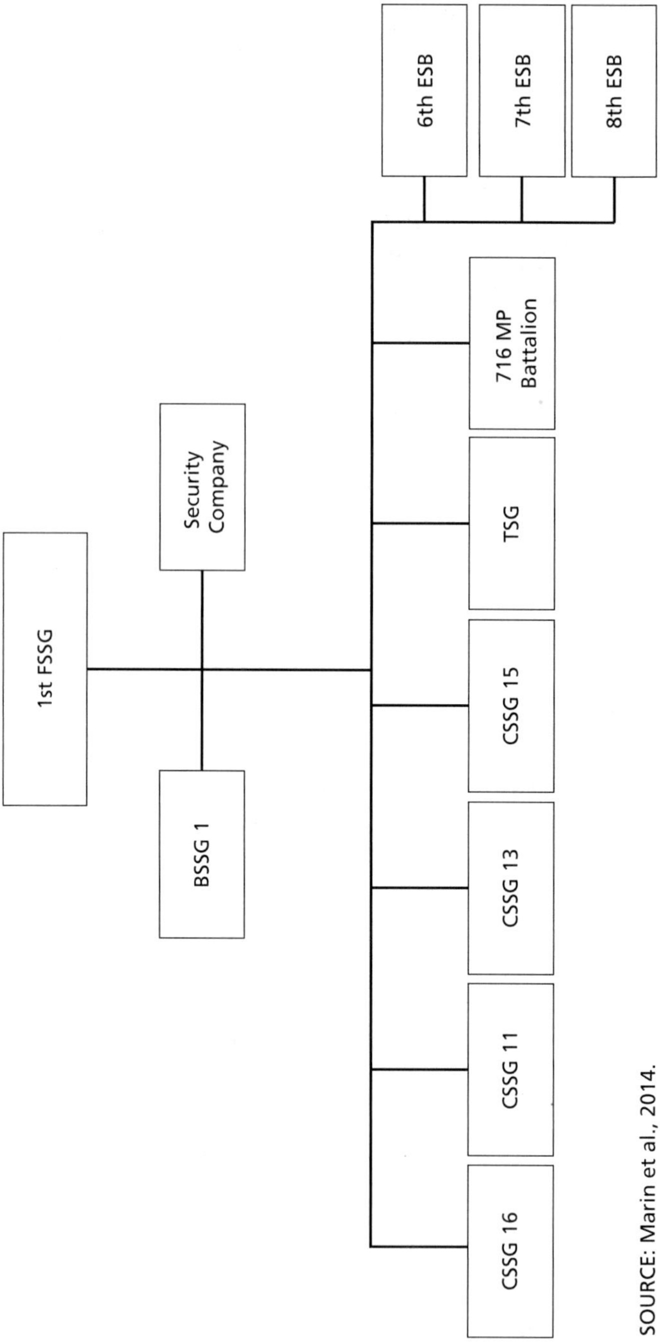

SOURCE: Marin et al., 2014.

RAND *RR1572-2.2*

landscape, the current structure has lost its economies of scale.[3] One development was the new growth in headquarter functions, at a loss of operators. During the reorganization, the MLG has gone from eight battalions to 12 battalions and three regiments. This has resulted in a large increase in the number of Marines filling headquarter roles at the expense of operational billets. Currently, more than 1,000 Marines fill headquarter roles. It is important to emphasize that this report is not an evaluation of whether the current LCE structure is sufficient to support its stated mission, but it seeks to demonstrate an approach to estimating logistics capacity that could serve as the foundation to further understand MLG capacity.

MLG Organization and Operations

The current MLG structure supports both DS and GS logistics missions. It is built around standing Combat Logistics Regiments (CLRs) and CLBs. Within 1st MLG, there is a headquarters regiment, a DS CLR, a GS CLR, an ESB, a medical battalion, and a dental battalion. Figure 2.3 illustrates the current structure for 1st MLG.

The headquarters regiment is responsible for a large percentage of the service functions of the MLG. It houses both a service company and a food service company. The headquarters regiment is responsible for providing C2 functions required to integrate services support. Subordinate units can be task-organized to provide support to smaller units. In addition to the service company and food service company, the headquarters regiment is task-organized with a communications company and the three CLBs that support the west coast MEUs. One of these MEU CLBs, CLB-11, is used in this report as an example of how LUEs can be applied to estimate logistics capacity.

CLR-1 is the DS regiment for 1st MLG. The mission of CLR-1 is to provide logistics support beyond 1st Marine Division's organic capability. It consists of a TSB and three DS CLBs with habitual rela-

[3] Robert Benbow and Patricia Neil, *MLG Reorganization: Focus and Analysis on the Headquarters Element*, Alexandria, Va.: Center for Naval Analyses, March 2007, p. 2.

Figure 2.3
Current 1st MLG Structure

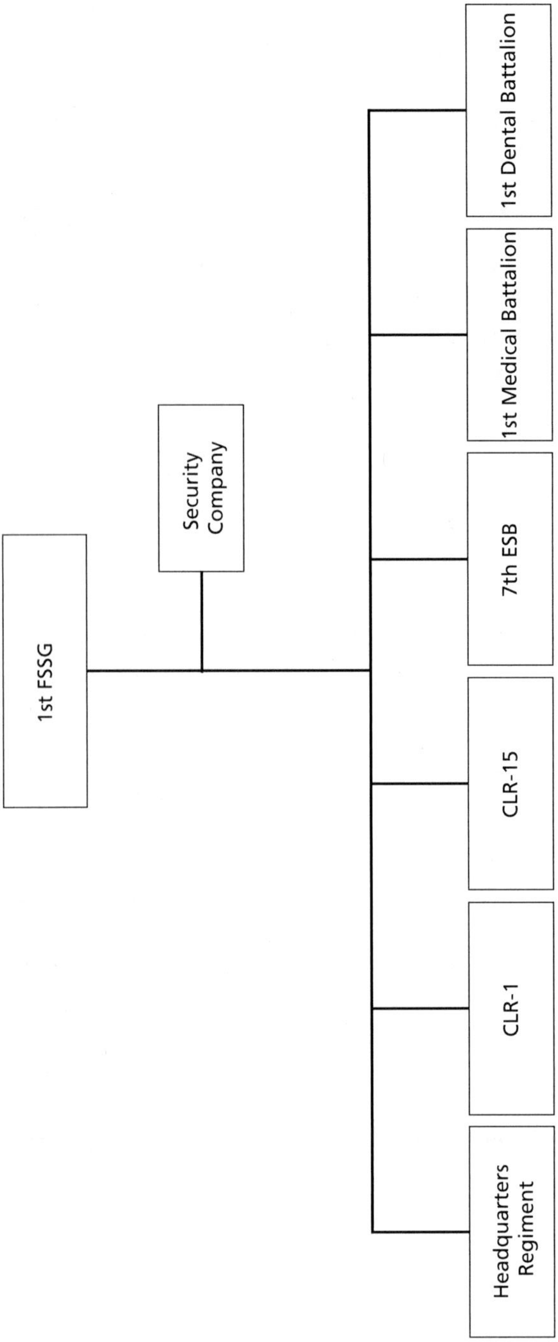

SOURCE: Headquarters, U.S. Marine Corps, 2015; and 1st Marine Logistics Group, 2015.

RAND *RR1572-2.3*

tionships with each of 1st Marine Division's three regiments. Each of the DS CLBs is task-organized to provide tactical logistics capabilities beyond the organic capability of the regimental logistics elements. These tasks include transportation, intermediate-level supply, field-level maintenance, and general engineering. The DS CLBs are organized differently than the MEU CLBs, with a greater focus on supporting the GCE rather than the entire MEU.

Within 1st MLG, GS logistics functions are conducted by CLR-15. Its mission is to provide intermediate-level supply, field-level maintenance, materiel distribution support, procurement management, equipment fielding support and forward resuscitative health care to the MEF. Within CLR-15, there are maintenance and supply battalions, as well as three Combat Logistics Companies. Supply and maintenance battalions form the bulk of CLR-15. The Combat Logistics Companies are small, task-organized companies that support Marine Corps Air Station (MCAS) Miramar, MCAS Yuma, and the Marine Corps Air Ground Combat Center Twentynine Palms. These companies are able to provide limited logistics capability across all six functions of logistics and primarily support large-scale exercises and training.

Separate from the DS and GS CLRs are the 7th ESB, 1st Medical Battalion, and 1st Dental Battalion. The 7th ESB provides general engineering support to the MEF. Its key tasks include: mobility, countermobility and survivability enhancements, explosive ordnance disposal (EOD), and general supply support to include the handling, storage, and distribution of bulk water and fuel. The 7th ESB contains certain elements not found elsewhere in the MEF, such as a bridging company. The 1st Medical Battalion provides health service support, and the 1st Dental Battalion is the source of dental services to the MEF.

In addition to the fixed structure, the MLG also provides LCEs to smaller MAGTFs, which can be permanently organized or task-organized for specific missions. One of the principle operating concepts of the Marine Corps in general is to generate MAGTFs that are task-organized, rapidly deployable, and self-sustainable. The logistics element is essential in ensuring these MAGTFs can accomplish these requirements.

The MLG is responsible for supporting both its own internal logistics and external service logistics requirements. As such, the MLG performs more than 150 tasks across the six primary warfighting functions. Specific to logistics, the MLG overall has eight METs, but each of the MLG's regiments and battalions also has METs. Table 2.1 identifies 1st MLG's core METs. Table 2.2 lists CLB-11's core METs and serves as an example of a battalion's METs. All of METs, including the core METs, number 75. However, removing the duplicates resulted in 47 unique MLG METs, which were analyzed for this study.

As seen in Table 2.2, CLB-11 lists nine core METs. However, seven of them are identical to 1st MLG's core METs. The two not already listed for 1st MLG, Provide Evacuation Control Center (ECC) and Facilitate Foreign Humanitarian Assistance, were added to 1st MLG's core METs as part of the 47 METs used to build LUEs.

The active-component MLGs are tasked to perform all six logistics functions. However, they are not task-organized to conduct two key logistics tasks—civil affairs and mortuary affairs. Units within the reserve component perform these two tasks.

Table 2.1
1st MLG's Core METs

MCT Number	Title
MCT 1.1.2	Provide Task-Organized Forces
MCT 4.1.2	Conduct Ground Supply Operations
MCT 4.2.2	Conduct Ground Equipment Maintenance
MCT 4.3	Conduct Transportation Operations
MCT 4.4	Conduct General Engineering Operations
MCT 4.5	Provide Health Services
MCT 4.6.1	Provide Logistics Combat Element (LCE) Support Services
MCT 6.8	Conduct Explosive Ordnance Disposal (EOD) Operations

Table 2.2
CLB-11's Core METs

MCT Number	Title
MCT 1.13.2.1	Provide Evacuation Control Center (ECC)
MCT 1.15.1.2	Facilitate Foreign Humanitarian Assistance
MCT 4.1.2	Conduct Ground Supply Operations
MCT 4.2.2	Conduct Ground Equipment Maintenance
MCT 4.3	Conduct Transportation Operations
MCT 4.4	Conduct General Engineering Operations
MCT 4.5	Provide Health Services
MCT 4.6.1	Provide Logistics Combat Element (LCE) Support Services
MCT 6.8	Conduct Explosive Ordnance Disposal (EOD) Operations

Impact of New Organization on Logistics Support

Under the FSSG structure prior to 2006, the Marine Corps employed a functionally aligned logistics structure that was capable of providing economies of scale to support the MEF. While it was able to support intermediate-level logistics requirements levied by the MEF, it lacked multifunctional units with habitual relationships between supported and supporting units. When the FSSG deployed, it was standard to create ad hoc organizations that possessed this multifunctionality. In 2006, the MLG was formed in order for the Marine Corps to "train like it fights." This led to the current MLG task organization that provides GS and DS tactical-level logistics.

This new task organization is closely aligned with key Marine Corps strategic concepts that call for smaller-scale forces that can expand rapidly while maintaining their organizational structure and modular packages with inherent flexibility. However, the Marine Corps continues to lack the ability to measure whether the MLG can meet operational logistics demands. For example, what is the transportation capacity of a DS CLB? How many missions can be supported under the current structure? These are all important questions whose answers help planners properly allocate assets.

CHAPTER THREE

Determining Logistics Capacity Metrics

This chapter presents a brief overview of the metrics used for logistics in other services and sectors, in order to determine what commonly-used metrics were of potential use to the Marine Corps. It also highlights perspectives within the Marine Corps logistics community, and the different levels of metrics that might be helpful in meeting various requirements among the community.

Marine Corps Stakeholder Perspectives

In order to develop metrics that were of value to the logistics community, we began with stakeholder interviews aimed at better understanding their desired goals. Discussions with stakeholders revealed that they were attempting to address different issues through the quantification of logistics capacity metrics and at different organizational levels. Even though the focus of this research effort was to create building-block LUEs, this helped identify the context for developing logistics capacity metrics and allowed the team to better understand how the Marine Corps might employ them.

We held discussions with the SAC members in September 2015. The purpose of these discussions was to better understand the types of logistics capacity metrics that would be the most useful to different stakeholders. In addition to 1st MLG, we were able to hear from Installations and Logistics (I&L), 2nd MLG, and Marine Forces Command (MARFORCOM). The 1st MLG provided additional information during the January 2016 visit to Camp Pendleton. Overall, different stakeholders have different levels of logistics capacity that they are most concerned with understanding and communicating.

Stakeholders who have a view of Marine Corps logistics from higher within the organization, such as I&L and MARFORCOM, appear to require more-aggregated capacity metrics. These stakeholders are interested in quantifying logistics capacity to better address such issues as readiness and force management. The MARFORCOM representative expressed a desire to understand the "residual capacity" of the MLGs to support additional missions after accounting for units that were already deployed.[1] Similarly, I&L expressed a "force provider perspective" that was interested in identifying the logistics force structure that was committed, and then understanding the capacity of what was left.[2] The context of the discussion with MARFORCOM was to understand and express the readiness of the MLG, and the context of the discussion with I&L was to ensure the correct force structure to meet future taskings.

On the other hand, stakeholders, such as the MLGs, expressed a desire to better communicate logistics capacity to GCE commanders. For instance, SMEs at 1st MLG were interested in being able to articulate the capacity of a certain logistics package and what that package would provide to the supported unit. The most appropriate logistics metrics in this case are higher fidelity than those needed for Marine Corps–level force management, and do best to demonstrate more-direct connections to specific tasks. A representative from 2nd MLG brought up the potential benefits of creating standardized logistics capability sets: avoiding confusion on the part of supported units and avoiding the "piece-mealing out" of logistics personnel in small units.[3] Individuals from 1st MLG noted some of the disadvantages of lacking standardized LCE structures, such as longer planning time and greater confusion on the part of the GCE on what support they were

[1] Telephone conversation with a representative from MARFORCOM G-4, September 21, 2015.

[2] Telephone discussion with a representative from I&L, September 29, 2015.

[3] Telephone discussion with a representative from 2nd MLG, September 17, 2015.

receiving.[4] The MLGs also discussed the importance of having logistics capacity metrics that were tied to the METs, as the METs directed training. 1st MLG also expressed the need to capture the demand for logistics support from the MEFs, beyond MEU CLB missions.

Based on these stakeholder objectives, different aggregate logistics metrics may be most relevant for different purposes. At a lower level within the organization, LUEs can serve to compare task-level capacity between logistics organizations. They can approximate standardized packages of logistics and aid in discussions between LCE and GCE commanders. (Although the stakeholder interviews tended to raise the example of communicating with GCE commanders, the MLG does more broadly support the MAGTF and the ability to communicate with all elements is important.) Therefore, the LUEs are closest to meeting the needs of the MLGs as capacity metrics in logistics. A logical next step is to turn the LUEs into a logistics estimation tool or dashboard that would demonstrate capacity based on current unit resources, which the Marine Corps is currently developing. Either further aggregation of the LUEs or a top-down heuristic separate from the LUEs is needed to create metrics with units that are useful to MARFORCOM and I&L for readiness and force management.

Review of Potential Metrics for Logistics Capacity

This section briefly reviews logistics capacity metrics from other services and domains. For the purpose of this report, *logistics capability* is defined as the ability to employ capacity competently to generate output from a logistics system over time. *Logistics capacity*, in turn, is defined as the measurement or estimate of the amount of output produced by a logistics system per unit time.[5] Therefore, an LUE constitutes a single-product logistics system—a system whose output is measured in terms of a single measure. An example would be repairs

---

4 In-person discussions at 1st MLG, January 8, 2016.

5 David S. Stoller, *Logistics Systems Capacity*, Santa Monica, Calif.: RAND Corporation, RM-4852-PR, 1966, p. 12.

completed, where no differentiation is made among types of repairs. This implies an average output, as certain repairs most certainly take more or less time. This requires an assumption of homogeneous units produced by each LUE. By measuring output, *capacity* is defined as the output of a single unit as determined by the manpower, equipment, and resources utilized. This output, dependent on the production rate, determines the sustained logistics activity that can be supported.

The increasing complexity of technology requires an ability to determine whether a logistics system has the required capability to perform the desired assignment. This often relies on the determination of the individual parts of a system's capacity. In most instances, *capacity* is defined as throughput; namely a quantity over time. Furthermore, "the ability to describe logistics in a purely methodological way allows for a more uniform assessment of the capabilities and for more effective solutions."[6] Such a method of determining logistics system capacity is conducted in various forms within both private industry and the military.

Both industry and the military have a common interest in understanding logistics capacity as it pertains to informing decisions to improve efficiencies and reduce costs. Most often these assessments of capacity are performance-based assessments. Private industry is primarily concerned with financial aspects when it comes to capacity—primarily how the system can reduce costs. Private industry is also often concerned with general indicators of a system's capacity.

However, the military is more concerned with other resources, such as equipment and personnel, and how logistics capacity affects readiness. They are also concerned with the specific capacity of given units, as that reflects what support the unit provides to its end user or supported unit. A key limitation to the military's approach of determining capacity is that while these capacities are sometimes captured, most often they are based on subject-matter expertise that resides with individuals. This can result in the loss of information if that person is reassigned or chooses to retire.

---

[6] Stoller, 1966, p. 12.

Capability is related to capacity. The models that the military uses to measure capability are varied and differ across the warfighting functions. The most common tool used by the military to measure capability is readiness ratings, which are inputted into the Defense Readiness Reporting System. This gives an assessment of a unit's training, equipment, and personnel readiness. For the variables that military readiness reporting systems measure, the methodology provides useful indicators, but it does not measure overall capacity of the system in terms of logistics output. While a logistics unit may meet its readiness requirements and demonstrate it can meet its tasks and missions, these readiness ratings do not accurately show how much the unit can provide the supported unit. Therefore, a way of measuring capacity is required at the building-block level to indicate the output of a given unit.

The Air Force and the Army have developed different models and tools that attempt to quantify logistics resources to help the services determine what mix of resources would maximize warfighting capability. The Army has been interested in developing a seamless logistics system that provides visibility of logistics resources, quantifiable information to inform battlefield decisionmaking, and effective support to the soldier—this is all part of its investment in better information and data to drive decisions. The systems the Army uses to achieve this are the Army Enterprise Systems Integration Program, Global Combat Support System-Army, and the Logistics Modernization Program. The defining metrics for these systems are those that support the individual soldier—what is needed or required to sustain the fight, sustain movement, and sustain combat power projection. These should be quantifiable metrics that lead to effective decisionmaking and are focused on the final end customer. However, most of the systems used by the other services are focused at the macro level and not at the micro level.

Levels of Metrics for Logistics Capacity

For the Marine Corps, different levels of logistics capacity metrics are more useful for different members of the stakeholder community. The levels of metrics build up from the bottom, with more-aggregate measures relying on the levels below. This is illustrated in Figure 3.1.

At the bottom of the pyramid is the planning-factor level of capacity metrics. Here, the units of measure are the number of passengers accommodated, the gallons produced, the number of requisitions handled, and other similar metrics. Metrics at this level are most closely associated with the operation of equipment or a productivity metric relating to individual or team output. However, these metrics are insufficient to communicate information to other elements of the MAGTF about what this means in terms of the ability to accomplish tasks. These metrics are also insufficient to understand what might go into standardized logistical packages. For these reasons, it is necessary to bundle information at this level into units that are able to accomplish those things.

This first aggregation of capacity metrics and mapping to identifiable tasks produces the LUE in blue in the middle of the pyramid in Figure 3.1. This mission-planning level of metrics does assist in communicating logistics support in terms of tasks, and gets to standardized estimates of T/O and T/E at the most basic and indivisible level. The main focus of this effort was creating these LUEs. However, the LUE level is still insufficient to answer many readiness questions, such as the residual capacity in the remaining MLG structure to support additional missions. To get to this purpose, it is necessary to aggregate the LUEs up one level by identifying different combinations of LUEs with each type of mission.

The top level of the pyramid in Figure 3.1 is about the number of supportable missions, with the focus on force management and readiness. The steps to aggregate to this level would be to: (1) identify the combinations of LUEs for different missions; (2) examine a given unit's available personnel and equipment to determine how many LUEs are available; and (3) see how many missions of what type are supportable by the unit's available supply of LUEs. In this report, we identified the starting point for support to HADR and NEO in terms of LUEs, but we would need additional data to complete an estimate on numbers of supportable missions.

Figure 3.1
Logistics Metrics Levels

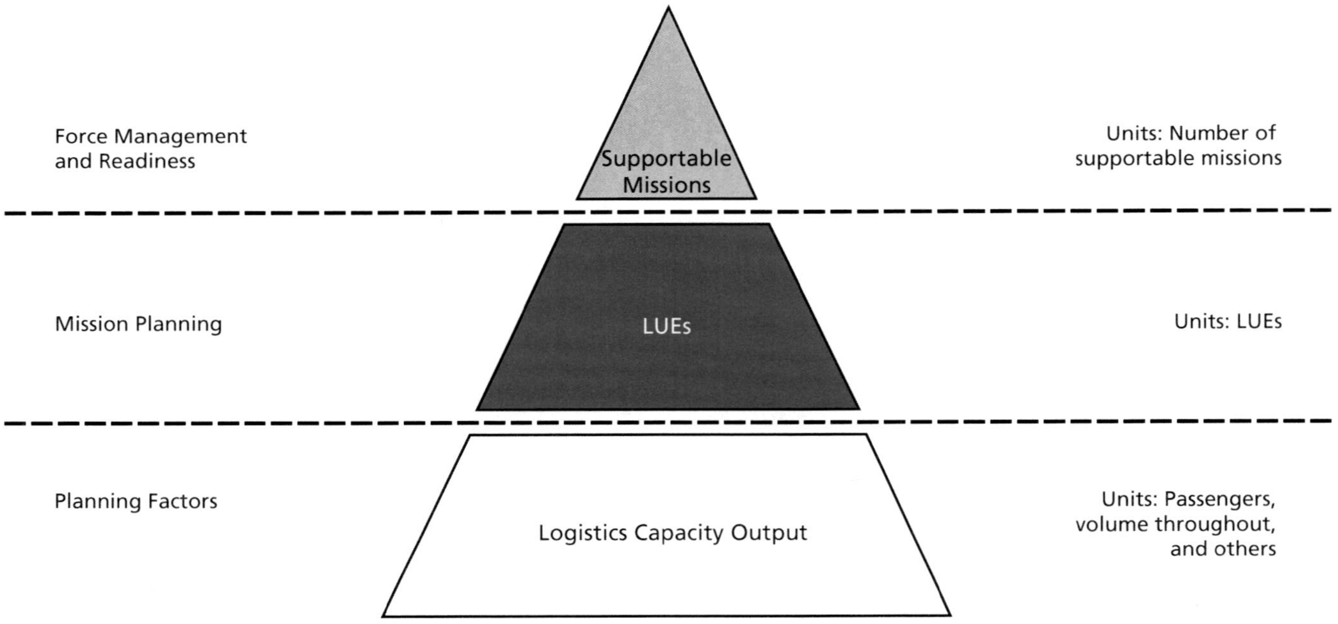

SOURCE: Headquarters, U.S. Marine Corps, 2015; and 1st Marine Logistics Group, 2015.

RAND *RR1572-3.1*

CHAPTER FOUR

Associating Mission-Essential Tasks to Logistics Units of Employment

It is important for operational and logistics planners to be able to effectively and clearly articulate the ability of a unit to provide logistics. This consists of providing well-defined terms and metrics to specifically state the output capacity of a logistics unit consisting of a set amount of equipment and personnel. It also requires that resources are available to ensure the unit is capable of effectively using its capacity to provide the needed logistics functions. This report offers the LUE as the building block required to perform a specific logistics task. The concept of an LUE is similar to a fire team in the GCE. That is, it is the smallest unit of employment and can be expanded or increased in order to provide greater logistics capacity. Just like three fire teams make up a squad, an LUE can be built from a truck section to a truck platoon or an EOD team to an EOD section. The LUEs are associated with METs and include personnel and equipment. Each LUE is associated with a metric that assesses the LUE's capacity to perform its task. Other key considerations for an LUE include the logistics unit's capability consisting of required training for personnel, rank requirements, C2 functions, and support equipment. However, within this report, these factors are not portrayed in the LUE.

In order to create LUEs, it is important to understand the core logistics capabilities of the MLG. *Logistics capability*, as defined by MCWP 4-1, *Logistics Operations*, is "the individual, functional logistics operating systems that exist at each level of war and are tied together by command and control."[1] At the tactical and operational levels, these consist of the capabilities required for the MLG to accomplish all six logistics functions. Therefore, LUEs are dependent and closely linked to the METs required of the MLG.

Methodology

To create LUEs, we reviewed the METL for 1st MLG and identified the smallest set of personnel and major equipment that would be needed to accomplish each of the required tasks. Available documentation and SME input factored into the LUEs. Although much of the information used to create the LUEs might already be well understood by the logistics community, the knowledge was scattered; now, however, the LUEs compile it in one location and for the different functions of logistics.

For certain tasks, such as Conduct Recovery and Evacuation Operations, multiple LUEs might be required. In that case, LUEs were created for wheeled vehicles, amphibious assault vehicles (AAVs), and tanks. For other tasks, a single LUE was sufficient. Once all key tasks were identified, a further review of Marine Corps warfighting doctrine was conducted. This was useful in identifying those building-block units that already existed. An example of this is a helicopter support team (HST). This is a team defined by doctrine with a defined task, clear organization, and identified equipment. Upon completion of doctrine review, LUEs were validated with SMEs during a field visit with 1st MLG. The end result was a list of LUEs that cover all key logistics tasks required of 1st MLG.

Demonstrating logistics capacity requires three key components: personnel, equipment, and appropriate metrics.[2] By combining these three essentials, capacity can be determined. For example, an MK27 7-ton truck with a driver and an assistant driver can transport 12.2 short tons (24,400 pounds) of cargo or 20 personnel. Therefore,

---

1 Headquarters, U.S. Marine Corps, 1999, pp. 1–2.

2 While availability of supplies is also a key component of the ability of personnel and equipment to move supplies, assume that supplies are available for this analysis.

the equipment (truck) and personnel (one licensed driver, one assistant driver) has the capacity to transport $x = 12.2$ short tons of things or $y = 20$ people. This can be scaled up to provide the aggregate capacity of a unit. If a platoon has 20 MK27s and 40 appropriate personnel, the unit has the capacity to transport $20x$ short tons or $20y$ personnel. However, in these cases, personnel or equipment may be limiting factors. Some units might have 20 trucks but not 20 licensed drivers. In that case, the maximum capacity of the unit would be dependent on the number of drivers. This is an overly simplistic example with clear metrics. As will be demonstrated later in the chapter, not all LUEs have as clearly defined capacity. The capacity of a warehouseman to draw supplies is perhaps more difficult to define. In these cases, an average figure is used for planning based on subject-matter expertise and historical data.

The next step once equipment, personnel, and metrics were assigned to each LUE was to review unit T/Os and T/Es. This process determines the number of LUEs that can be fielded by each unit. To do this, numbers of principle end items (PEI) and appropriate military occupational specialties (MOSs) are assigned to LUEs. A key challenge with this approach is that many pieces of equipment are dual use or alternate pieces of equipment could be used in lieu of the primary PEI. For instance, an ambulance team can transport patients via either a two-litter ambulance or a four-litter ambulance. Each piece of equipment is capable of completing the required task, Conduct Casualty Evacuation. However, each has a different capacity. Because of time constraints, this validation was conducted for a single unit, CLB-11.

Limitations

The LUEs presented in this study serve the purpose of measuring capacity. As such, they consist of personnel, equipment, and metrics and do not take personnel qualifications into account. In order to perform certain tasks, logistics personnel should be of a certain rank, possess certain training, or have obtained certain licenses. However, the RAND team assessed these details to be measures of capability and beyond the scope of the study to articulate this level of detail. The Marine Corps attempts to fill billets with personnel who fit these requirements, but sometimes it is forced to fill with lesser ranks or a unit has trouble sending the Marine to training. For example, in order to conduct en route care, the nurse assigned to the LUE needs to have attended en route care training. This is a school with limited seats, and the Navy school is going away. Continuing to have access to personnel with appropriate training will be a challenge. Units must conduct appropriate troop-to-task assessments of their units to determine whether they have the appropriate personnel to meet the LUE requirements and demonstrate that LUE's capacity.

Similarly, the LUEs contain limited ability to demonstrate the substitutability of personnel and equipment: What other types of personnel or equipment are close substitutes for the ones listed in the LUE, if the preferred capabilities are not available? Again, this is a measure of capability and not capacity. The LUEs for the most part contain the optimal pieces of equipment and personnel preferred by doctrine and SMEs. However, the LUEs do provide capacity information for alternate pieces of equipment that reside within the MLG.

Functions of Logistics

The next several sections record a description of the METs for 1st MLG by the six logistics functions. Subfunctions within each logistics function are also discussed where appropriate.

Transportation

Transportation operations consist of distribution of supplies across various means, such as railways, highways, waterways, and airways. The key subfunctions of transportation operations are motor transport (MT), materiel handling, landing support, embarkation, freight/passenger transportation, aerial delivery, and port and terminal operations. A list of the transportation METs that the MLG is required to perform is included in Table 4.1.

Table 4.1
Transportation Mission-Essential Tasks—1st MLG

MCT Number	Title
MCT 4.3	Conduct Transportation Operations
MCT 4.3.2	Conduct Port and Terminal Support
MCT 4.3.3	Conduct Motor Transport Operations
MCT 4.3.4	Conduct Air Delivery
MCT 4.3.6	Conduct Materiel Handling Operations
MCT 4.3.9	Conduct Landing Support Operations

Transportation operations are spread out across multiple units within the MLG and are organized to provide both DS and GS capabilities. The TSB primarily provides GS transportation services to the MEF while the numbered CLBs (1, 5, and 7) provide transportation support to their supported infantry regiments beyond their organic capacity.

Motor Transport Operations

MT is perhaps the most versatile of the Marine Corps transportation tasks, and it consists of providing surface transportation by wheeled vehicles. The MLG is responsible for providing both tactical and logistical transport support to elements of the MAGTF. This can be done through GS, which involves medium- and heavy-lift support for the MAGTF, or through DS for the Combat Service Support Element (CSSE). Personnel, cargo, and bulk liquid are the items that the MLG is tasked with transporting. The primary unit of employment to conduct MT operations is an MT convoy team. These are often task-organized based on mission requirements; however, each convoy team consists of a mix of the following components:

- convoy C2
- security element
- cargo section
- bulk liquid assets

- personnel transport section.

Depending on the mission, some of these components can be omitted. The size of the team is dependent on the amount of items that require transportation, the time allotted to transport, route length, and enemy situation.

In order to determine a standard for the convoy team LUE, the team referred to the T&R standards, which define the capability to the platoon size level. An MT platoon has the capacity to support an infantry company, while an MT company supports an entire battalion. In recent years, the Marine Corps MT community has begun referring to convoy teams as Combat Logistics Patrols (CLPs). A CLP on average consists of 11 vehicles. This includes one wrecker, a fuel truck, a water truck, and eight cargo Medium Tactical Vehicle Replacements (usually a mix of short- and long-bed trucks). This allows a platoon of trucks to deliver 7,200 gallons of fuel, 7,200 gallons of water, six to eight Twenty-foot Equivalent Units of cargo, and vehicle recovery capability.

While the CLP is the unit of employment used by MT units, it is useful to break down the CLP into smaller equipment components to determine the capacity of the unit. For the purpose of determining capacity for motor transportation, we created four variations of the motor transport team (MTT) LUE:

- MTT–personnel
- MTT–bulk cargo
- MTT–water distribution
- MTT–fuel distribution.

Each of these teams consists of one piece of equipment capable of accomplishing the subtask and the metric associated with it. For instance, to move personnel, a logistician could employ an MK23, which can move 18 personnel, or they could employ an MK27, which has the capacity to move 20 personnel. An MT operator (MOS 3531) is capable of operating either of these pieces of equipment and is therefore interchangeable. The MK23 and MK27 used in the previous example could also be used to transport bulk cargo. The capacity of each piece of equipment within the MLG is therefore also articulated for that LUE. The High Mobility Multipurpose Wheeled Vehicles included in the MTTs do not require an MT operator, rather they can be employed by any Marine with an incidental driver's license. In addition, trailers add additional capacity and are incorporated in the bulk cargo LUE. Therefore, to demonstrate unit capacity, a logistician would determine how many assets he or she possesses and multiply the capacity by the number of assets as long as the logistician has enough operators for every vehicle.

Materiel Handling Operations

Materiel handling is a high-demand, critical task provided by logistics. It is required for both internal sustainment and external logistics support. Effectively employing materiel-handling equipment (MHE) is essential to the throughput of equipment. A materiel-handling LUE consists of the MHE required and the operator (MOS 1345). A single piece of equipment constitutes an LUE because MHE assets are often employed individually. The MHE resident in the MLG has the capacity of providing lift up to 7.5 tons. While pounds lifted is a useful capacity metric for the individual piece of equipment, another key metric useful to the commander or logistics planner is the number of sites supported. Therefore, the number of sites supported is also used as the metric associated with the MHE LUE. The number of sites supported is dependent on time and space. If sites are closely located and tasks can be conducted sequentially, then one MHE LUE will suffice. However, if the sites are farther apart or MHE is required simultaneously, then one LUE may not be appropriate.

Landing Support Operations

During the ship-to-shore movement of an amphibious operation, landing support provides assistance in effectively moving people, supplies, and equipment across the beach to continue operations ashore. The primary subtasks of landing support are to provide the following groups/teams:

- Arrival/Departure Airfield Control Group (A/DACG)
- Beach Operations Group (BOG)

- HST
- Port Operations Group (POG)
- Rail Head Operations Group (RHOG).

All of these units provide throughput services to move personnel, supplies, and equipment ashore and marshal for follow-on movement.

Landing Support (LS) Company is where the majority of LS assets reside within the MLG. LS Company consists of two LS platoons; each platoon has the capacity of supporting a BOG, POG, RHOG, or A/DACG. Therefore, TSB, the battalion headquarters for LS Company, can support one of each of these missions simultaneously along with three HSTs. However, upon further review of MLG's support to SPMAGTFs and the MEU, it was determined that the platoon could be further subdivided into smaller components to demonstrate capacity. Teams of nine to ten LS Marines (MOS 0481) are usually employed in smaller MAGTFs. These teams are capable of providing a more limited output across all of the aforementioned missions. For instance, an LS platoon can support a colored beach made up of three numbered beaches for a BOG.[3] The smaller LS team could support simply one numbered beach. The primary metric for determining LS capacity for each of these LUEs is throughput. However, for the commander, the number of beaches and sites supported is also useful.

Most of the LS LUEs are already defined in Marine Corps doctrine, making it an area where determining unit capacity is much easier. For instance, an HST as defined by doctrine has a clearly defined task organization of eight to ten LS Marines with defined roles and responsibilities augmented by support personnel:

> (1) HST commander (has to be at least a Sgt.); (1) radio operator; (1) corpsman; (1) safety NCO [noncommissioned officer];

[3] According to joint doctrine, a *colored beach* is the portion of usable coastline sufficient for the assault landing of a regimental landing team or similar sized unit; see Joint Staff, U.S. Department of Defense, Joint Publication 3-02, *Amphibious Operations*, Washington, D.C., July 18, 2014b.

> (1) hookup man; (1) static man; (1) inside director; (1) outside director; add up to 4 legmen (2 for single point; 4 for dual point).[4]

This HST organization is able to conduct a single point lift and can lift up to 40,000 pounds of cargo with the appropriate nets and slings.

Port and Terminal Support

The term *port operations* refers specifically to the loading and offloading of ships. *Terminal operations* on the other hand consist of the "reception, processing, and staging of passengers; the receipt, transit, storage, and marshaling of cargo; the loading and unloading of ships or aircraft; and the manifesting and forwarding of cargo and passengers to destination."[5] Terminal operations apply to all manner of transportation. To conduct port and terminal support, the same LUEs are required as in landing support operations.

Air Delivery

Air delivery is the rigging of equipment and supplies to provide in-flight delivery to troops on the ground. All air delivery capability resides in the LS Company of the TSB. It is important to note that the air delivery platoon is not capable of supporting itself and requires support from the TSB for such things as MHE support. The primary LUE for conducting air delivery operations is an air delivery team. It consists of four Aerial Delivery Specialists (MOS 0451), and it has the capacity—with the appropriate equipment—to support airdrops for an infantry rifle company. While the weight of the airdrop (in pounds) is the primary metric used to articulate air delivery capacity, other metrics that describe what is being dropped (e.g., ten days of supply for infantry company) can also be useful.

General Engineering

Engineering personnel and assets within the MLG are primarily within the ESB. The mission of the ESB is to "provide expeditionary verti-

---

[4] Headquarters, U.S. Marine Corps, 2001a.

[5] Headquarters, U.S. Marine Corps, 2001a, 1-3.

cal and horizontal construction, standard and nonstandard bridging, EOD support, bulk fuel operations, bulk water operations, and tactical utility support to the MAGTF."[6] The ESB can be employed in GS of the MAGTF or provide task-organized units. Creating engineering teams capable of accomplishing each of their core tasks primarily does this. While engineering assets reside within other units in the MLG, most of these assets are employed to sustain the internal requirements of the unit (e.g., utilities support) and these units do not provide engineering support to external units. The one exception to this is the multifunctional MEU CLBs. Engineering METs are listed in Table 4.2.

Table 4.2
General Engineering Mission-Essential Tasks—1st MLG

MCT Number	Title
MCT 1.4.1	Conduct Mobility Operations
MCT 1.4.2	Conduct Counter-Mobility Operations
MCT 2.2.2	Provide and Maintain Engineering Reconnaissance Operations
MCT 4.4	Conduct General Engineering Operations
MCT 4.4.1	Conduct Horizontal Construction
MCT 4.4.1.1	Conduct Limited Horizontal Construction
MCT 4.4.2	Conduct Vertical Construction
MCT 4.4.2.1	Conduct Limited Vertical Construction
MCT 4.4.3	Conduct Bulk Liquid Operations
MCT 4.4.3.1	Conduct Limited Bulk Liquid Operations
MCT 4.4.4	Conduct Tactical Electrical Supply
MCT 4.4.4.1	Conduct Limited Tactical Electrical Supply
MCT 6.8	Conduct Explosive Ordnance Disposal (EOD) Operations

[6] Headquarters, U.S. Marine Corps, 2000a, p. 1-9.

Vertical and Horizontal Construction

Vertical and horizontal construction are the primary tasks of combat engineers (MOS 1371). Vertical construction is the building or improvement of existing structures for use by the MAGTF. Horizontal construction shapes the terrain, such as road building and maintenance, airfield construction, and site preparation. These tasks are often completed by small engineering teams and can be scaled up to accomplish larger projects. For instance, a ten-person construction engineer squad can be employed to build a project of a certain size. As project demands increase, the squad would be built up to include additional squads. A platoon, consisting of three squads, could therefore create a larger vertical construction project, such as a Combat Outpost (COP). For the purpose of construction engineering, the engineering squad is the base unit according to multiple SMEs. The squad can be broken down to the team level consisting of three to four combat engineers (MOS 1371), but the project would be a small task. For the purpose of LUEs, the squad serves as the building block for determining construction capacity.

Determining meaningful metrics for construction projects can be difficult and subjective. Multiple construction metrics can be used, such as structures built per day, cubic yards of earth moved per hour, or linear feet cleared per hour. We interviewed one SME who provided an example of the capacity of a 90-person ESB engineer line company. In his experience, the company constructed a 300-meter–by–300-meter forward operating base (FOB) in 72 hours. This included the placement of 2,000 meters of concertina wire, construction of five guard towers, berms, three below-ground security bunkers, and an expeditionary helicopter landing zone. Upon final construction, the FOB consisted of a 22.25-acre facility that could accommodate approximately 1,000 personnel. Taking this example, the project size was deemed the most effective metric that would have meaning for the supported commander. A squad could conduct a SWA hut-sized project, a platoon could build a COP, and a company could build a FOB. Additionally, delineation of key tasks that could be completed by units would be of use to the supported commander. Conditions that could potentially affect the ability to achieve these capacity metrics include (but are not limited to) time, environment, enemy situation, and materiel availability.

Tactical electrical supply

These tasks include providing mobile electric power and cooling services. These services can be provided at the MAGTF level of support or in a limited manner. Tactical electric supply provides power generation and distribution through the Mobile Electrical Power Distribution System (MEPDS). For limited tactical electric supply, power generation is used to sustain the unit providing power.

The engineer support company within the ESB provides tactical electrical supply for the MEF, whereas a utilities section within other units would provide the organic limited tactical supply required. The primary assets used to provide power are the family of Marine Corps generators and their associated MEPDS equipment. The family of generators operates at 60 hertz and ranges from 2 kilowatts (kWs) to 100 kWs power. Electricians (MOS 1141), deployed in pairs to allow for 24-hour operations, manage the generators. According to the Marine Corps standard for electrical supply, the primary metric for determining power capacity is kW power produced daily. While this is an easy metric to assess, discussions with SMEs indicated that this metric is not useful to the supported commander. In that case, metrics that demonstrate the number of things powered are potentially more useful. For instance, two 802 generators produce 10 kW power a day when used in tandem; however, more importantly to the supported commander, they have the capacity to provide power to a company size Combat Operations Center (COC). Obviously, power requirements are determined on a case-by-case basis once a unit establishes its footprint, but understanding the building block capacity assists logistics planners in allocating appropriate power resources.

Similar to electrical supply, utilities Marines are also responsible for providing cooling services. The primary pieces of equipment used by the Marine Corps are the B0008 and B0014 air conditioning units. These two pieces of equipment provide 36,000 British Thermal Units (BTUs) and 60,000 BTUs of cooling, respectively. Management of this equipment is conducted by refrigeration/air conditioning technicians (MOS 1161). The planning ratio is one technician per every five pieces of equipment. Similar to power supply, the current metric of BTUs demonstrates capacity, but it is a difficult concept for the supported commander to interpret. Metrics that demonstrate size of structure cooled are potentially more effective.

Bridging

The only bridging assets in the entire MEF are located in the bridge company within the ESB. The mission of the company is to provide standard bridging and ferrying support, which are essential to the movement of the MAGTF. The bridging company provides two types of bridging assets—tactical bridges and floating bridges. These are the two primary units of employment.

A tactical bridge requires a platoon of 35 engineers to employ and is augmented by support Marines such as heavy equipment operators and MT operators. One platoon is capable of erecting one medium girder bridge (MGB) that is 90 feet in length with a military load capacity of 70. In order to erect another MGB or a larger MGB, the number of engineers remains constant, while the amount of equipment increases. For a floating bridge, the smallest unit of employment would be a squad of nine engineers augmented with MT operators. A squad of this size has the capacity of running one raft bridge of 66.5 feet. In order to run a continuous span floating bridge a platoon of 35 engineers would be required.

Bulk Liquid Operations

Bulk liquid tasks are broken down into two key levels of tasking—bulk liquid support and limited bulk liquid support. Bulk liquid support provides bulk fuel and bulk water for the MAGTF to support operations in forward deployed environments. This is done through use of the Amphibious Assault Fuel System and Tactical Water Distribution System capabilities. Other bulk liquid operations that support the MAGTF include shower services and laundry services.

Limited bulk liquid support consists of scaled-level engineer point services that are limited by the capacity of the unit. The primary planning factor for limited bulk liquid operations is fuel storage capacity less than 40,000 gallons and water storage capacity less than 10,000 gallons. In this case, bulk liquid is usually distributed from a single point via the Expedient Refueling System, Lightweight Water Purification System, and SIXCON modules.

The primary metric for measuring bulk liquid capacity is gallons per unit of time. For water and fuel distribution, it is specifically gallons per hour. Often the rate of water and fuel distribution is dependent on the pump module used in conjunction with the distribution system. Water and fuel storage is measured in gallons, as well as number of sites a team has the capacity to support. While there is no specific time, storage is most often accounted for by 24-hour period. Laundry services are measured in pounds of laundry per hour and shower support by the number of shower points and number of Marines serviced per hour.

Explosive Ordnance Disposal Operations

Most EOD assets are within the EOD Company at the ESB; however, there are other EOD assets elsewhere in the MLG, such as within all three MEU CLBs. EOD Company is tasked with the "detection, identification, recovery, evacuation, and disposal of items of unexploded ordnance."[7] An EOD company is made up of four EOD platoons and doctrinally supports the MEF. A platoon consists of three sections of eight EOD technicians and supports an infantry regiment.[8] Each section can support an infantry battalion. The building block component for EOD is the three-man response element, which can be further task-organized to a two-man team. This unit allows EOD to support distributed operations. The number of response elements depends on the number of sites supported, as well as the threat level and required mission.

To build the EOD LUE, we began with MCT 6.8, Conduct EOD Operations (listed in Table 4.2). We determined from the available documentation and SME discussions that the smallest, indivisible set of personnel and major equipment required to conduct EOD operations is an EOD section plus appropriate equipment. This translates to three EOD technicians (MOS 2336) and the following equipment:

---

7 Headquarters, U.S. Marine Corps, 2000a, p. 1–9.

8 Headquarters, U.S. Marine Corps, 2000a.

one MRAP, one VRC-110, one Talon EOD robot, one PacBot EOD robot, one .50 caliber dearmer kit (containing firing devices and fuses), one search kit, one demolition kit, one disposal kit, one improvised explosive device kit, and three tech kits. One such EOD LUE can conduct EOD operations at one site.

Supply

Two areas of logistics proved to be especially difficult to understand from a capacity perspective: supply and maintenance. In the supply community, there is the distinction between *direct support* and *general support*. While the capacity needed for DS to a specific unit may be easier to capture, it is more difficult to estimate how the GS capabilities that the MLG needs to maintain to support a variety of tasks and units should be measured. To determine capacity of supply tasks, a slightly different approach was taken. Table 4.3 lists the supply METs.

We first identified DS tasks. This includes limited administrative functions: fiscal, due and status file management, consolidated memorandum receipt management, and basic warehousing. In the case of the MEU, this might also include the ability to manage a class IX parts block and a float block. To identify a supply unit capable of accomplishing these tasks, the team looked at past manning documents for recent MEU and SPMAGTF deployments. From there, looking at the

Table 4.3
Supply Mission-Essential Tasks—1st MLG

MCT Number	Title
MCT 4.1.1.3	Conduct Reparables Management
MCT 4.1.1.6	Conduct Supply Chain Materiel Management
MCT 4.1.2	Conduct Ground Supply Operations
MCT 4.1.2.1	Determine Requirements
MCT 4.1.2.2	Conduct Procurement
MCT 4.1.2.3	Provide Storage
MCT 4.1.2.4	Conduct Distribution Operations

average number of requisitions and with the help of SMEs, we were able to identify a building block supply detachment that could accomplish the basic DS supply tasks. The supply detachment consists of one supply officer, four supply administrative Marines, and three warehousemen. The output of a team this size is an average of 125 requisitions a month. Other DS supply tasks include providing an ammunition supply point (ASP) and expediting. These teams are already fairly well defined within the supply community and so the LUEs for these two tasks reflect the average composition.

To determine capacity for GS supply tasks, we took a top-down approach rather than a bottom-up approach. Looking at the task organization of each of the sections within the supply battalion (the unit within MLG tasked with providing general supply support), we worked with SMEs to determine where the redlines in personnel were for each section in order to conduct their basic tasks. For instance, in order for the Deployment Support Unit to be able to meet its mission, it needs to be able to expedite, conduct materiel handling, manage the generator package, conduct quality control, and provide assistance in building and managing deployment blocks for exercises and deployments. While the current task organization of 62 is able to execute these tasks, there are built-in redundancies to accomplish the required tasks multiple times over. Therefore, upon further review of the task organization, the redline seems to be approximately 18 Marines—one supply officer and 17 supply administrative clerks. This would be the minimum personnel required for the mission. A section of this size has the capacity to support 15 deployment blocks for both training and operations. This same method was repeated for the other general supply support LUEs to include: general accounting, initial issue provisioning, fiscal, property accounting, storage, packing, customer service, rations, medical logistics, and ammunition.

Maintenance

In order to determine maintenance capacity, an evaluation of METs provided limited assistance. This is due to the fact that most maintenance units are multifunctional in their ability to conduct tasks. Rather than create teams based on tasks (e.g., testing and calibration, inspection, and classification), maintenance is broken down by type of equipment serviced. Therefore, determining capacity to repair MT equipment, engineering equipment, or communications gear is a more useful delineation.

Marine Corps maintenance capability is divided into two LOMs: field and depot. Field-level maintenance is divided into two sublevels: organizational and intermediate maintenance. Organizational maintenance is the responsibility of the using unit. Intermediate maintenance requires a higher level of technical proficiency. A similar approach to developing maintenance LUEs was taken, as demonstrated in the supply section earlier in this chapter. To determine organizational maintenance LUEs, we identified the required tasks at that echelon. From there, we developed task organizations. For intermediate maintenance, a somewhat similar redline approach was used. We, along with maintenance SMEs, looked at the task organization for each of the maintenance companies within the maintenance battalion and identified how many detachments of organizational maintenance sections the company could source before it was unable to complete its assigned mission. This was the redline used to determine the size of the intermediate maintenance section. Table 4.4 records the maintenance METs.

Depot-level maintenance involves tasks beyond field maintenance. Examples of depot-level tasks are inspection, repair, overhaul, modification, or reclamation of weapons systems, equipment end items, parts, components, assemblies, and subassemblies.[9] Depot-level maintenance was outside the scope of this present effort, but would also be an important logistics capacity to capture.

Organizational maintenance capabilities are divided into vehicle recovery teams and maintenance contact teams. Vehicle recovery teams consist of the vehicle recovery asset (e.g. wrecker for wheeled vehicles, R7 for AAV recovery), the vehicle operators, and the associated mechanics. The primary metric associated with these LUEs is tons (lift and tow). Maintenance contact teams are created for each of the equipment types already identified (engineering, MT, communications/electronics, and EOD). A contact team consists of two mechanics, and each mechanic

[9] Headquarters, U.S. Marine Corps, 2014a, pp. 16–17.

Table 4.4
Maintenance Mission-Essential Tasks—1st MLG

MCT Number	Title
MCT 4.2.2	Conduct Ground Equipment Maintenance
MCT 4.2.2.1	Conduct Inspection and Classification
MCT 4.2.2.2	Conduct Service, Adjustment, and Tuning
MCT 4.2.2.3	Conduct Testing and Calibration
MCT 4.2.2.4	Conduct Repair
MCT 4.2.2.5	Conduct Modification
MCT 4.2.2.8	Conduct Recovery and Evacuation Operations

has the capacity to maintain five pieces of equipment. As already mentioned in Chapter Three, this assumes homogeneity for repairs. All repairs are roughly equivalent. The primary equipment associated with maintenance contact teams are the required toolkits and a maintenance contact truck.

Health Services

The MLG provides both internal medical support and external medical support to MAGTF elements. The MLG provides both Level 1 and Level 2 medical care. The medical battalion within the MLG is the only unit within the MEF that provides Level 2 care. It is currently task-organized into building-block units, similar to an LUE, in order to provide specific medical services. Each surgical company shares a similar task organization and is broken into further subcomponents to include

- Forward Resuscitative Surgical Section (FRSS)
- Shock Trauma Platoon (STP)
- En Route Care Service (ERCS)
- holding ward
- lab section
- radiology section
- ambulance section

- dental section (augmented by a dental battalion).

The MLG also supports Level 1 medical care with battalion aid stations (BASs) and group aid stations. Table 4.5 lists the health services METs.

Forward Resuscitative Surgical Section

The FRSS is responsible for providing trauma care to casualties. They are the primary unit of employment for delivering resuscitative care. There are eight members of an FRSS—two surgeons, an anesthesiologist, a critical care nurse, a surface independent duty corpsman (IDC), medical corpsman, and two surgical technicians. Equipped with one Authorized Medical Allowance List (AMAL) 645 and five AMAL 646s, the team can provide care to 18 casualties in a 48-hour period. At any one time, FRSS can accommodate five patients (two preoperatives, one intraoperative, and two postoperatives) and has a maximum patient-holding capacity of four hours.

Shock Trauma Platoon

The task organization for the STP is 18 personnel and one AMAL equipment block (631) and two AMAL consumable blocks (632). The 18-person team consists of two emergency medical specialists, a physician's assistant, a surface IDC, an emergency room nurse, and 13 hos-

Table 4.5
Health Services Mission-Essential Tasks—1st MLG

MCT Number	Title
MCT 4.5	Provide Health Services
MCT 4.5.3	Conduct Casualty Treatment
MCT 4.5.4	Conduct Temporary Casualty Holding
MCT 4.5.5	Conduct Casualty Evacuation
MCT 4.5.7.2	Maintain Dental Health Readiness
MCT 4.5.7.3	Provide Emergency Dental Services
MCT 4.5.8	Conduct Medical Regulating

pital corpsmen. Their function is to provide medical support to a medical element, including triage, communication, security, and patient movement. It is most commonly employed with either a BAS or an FRSS. An STP with this task organization has the capacity to support 50 casualties.

En Route Care Service

An ERCS consists of one en route care nurse and one en route care trained corpsman. Equipped with an AMAL 647, this team can provide en route care for two critically injured, but stable casualties for up to two hours during flight. They are employed on medium and heavy lift airframes such as the MV-22 and CH-53 aircraft.

Holding Ward

The holding ward section has a ten–flow-through bed capacity. It is capable of providing 100 bed days with an average patient requiring 72 hours of bed rest. The section is staffed with a medical corps officer, two registered nurses, and eight hospital corpsmen. The AMALs associated with the holding ward are the AMAL 633 acute care ward equipment block and the AMAL 634 acute care ward consumables.

Lab Section

A lab section is made up of three medical lab technicians and the AMAL 618 lab equipment and two AMAL 619s lab consumables. A lab section has the capacity of performing hematology, microbiology, urinalysis, and chemistry testing for 100 patients.

Radiology

The radiology section is responsible for establishing an X-ray suite. The section is manned with three advanced X-ray technicians. They can establish one X-ray suite with the AMAL 627 X-ray equipment.

Ambulance Section

Ambulance sections are employed to evacuate casualties. An ambulance section consists of two ambulances, either an M997 (capacity to move four litter or eight ambulatory patients) or an M1035 (capacity to move two litter or four ambulatory patients). The section is manned with two ambulance drivers and two field medical technicians. Depending on which ambulance is used, the section has the capacity to move four to eight litter patients and eight to 16 ambulatory patients.

Dental Section

A dental battalion provides task-organized dental sections that provide operational dental support. The smallest task-organized dental formation is a field dental clinic and it is staffed with one dentist and one dental technician. Equipped with the Authorized Dental Allowance List (ADAL) 662 field dental operatory, this ADAL can service up to 2,200 patients before requiring resupply. The planning factor for a steady stream of patients is that a dental section can provide care to a defined patient stream for one month (175 dental casualties is the planning estimate).

Battalion Aid Station

A BAS consists of one medical officer, a surface IDC, and nine corpsmen. The equipment sets for the BAS are the BAS AMALs (equipment and consumables), as well as a sick call block. The number of patients is the metric to measure capacity, and a team of this size can provide medical support to 50 casualties, but it is limited to holding for only six hours.

Services

Services consist of both CSS and CSSE services.[10] CSS services are those inherent to a command, such as personnel administration; religious ministry; billeting; financial management; morale, welfare, and recreation; and messing. On the other hand, CSSE services are those services not available or not organic to other MAGTF elements, including postal, legal, field exchange, disbursing, mortuary affairs, and civil-military operations (CMO). The last two, mortuary affairs and civil-military operations, are not provided by the active-duty MLG; rather, they are provided by Marine Corps Reserve units and will not be included in this analysis. We did not include Operational Contract Support (OCS), a service recognized in Joint doctrine. OCS would be an area for further potential development of the LUEs.

---

[10] Headquarters, U.S. Marine Corps, 2001b.

For the purpose of this report, we are interested in measuring logistics capacity; therefore, we will focus on CSSE services. However, there is one exception. Messing is a logistical function, and it will be included in this section to measure capacity even though it is a command support service. In garrison, units providing CSSE services functions reside within the services company of the Headquarters Regiment. While they are centralized in garrison, their primary employment when deployed is in task-organized detachments to support the MAGTF. Most of these teams are singly focused by function as described in the following section. However, there is one unit of employment that is multifunctional and often employed in a deployed environment: the Warfighter Express Services (WES) team. It is made up of two postal clerks, one disbursing agent, and three Marine Corps Community Services (MCCS) Marines. The team provides services to roughly 100 to 300 Marines. Table 4.6 lists the services MET.

Postal

Postal, like most services functions, task-organizes its support into detachments. The primary units of employment are a mobile main postal office and smaller mobile-unit post offices. A mobile-unit post office is the building-block LUE and has the capacity to provide postal support to a reinforced regiment. Its task organization consists of two postal clerks (MOS 0161) and one clerk has the capacity to serve 500 Marines. The planning factor for pounds of mail is 1.75 pounds of mail per each Marine. Therefore, a mobile-unit post office can serve 1,000 Marines with an estimated maximum daily capacity of 1,750 pounds of mail.

Table 4.6
Services Mission-Essential Task—1st MLG

MCT Number	Title
MCT 4.6.1	Provide Logistics Combat Element (LCE) Support Services

Legal

Legal services encompass a wide variety of tasks, including command advice, military justice, legal assistance, operational law, fiscal law, and detention operations. Legal service support teams are task-organized detachments that provide legal services. Their composition is dependent on mission, size of supported MAGTF element, and expected duration. However, a standard unit of employment consists of one attorney and one clerk. This basic unit can be expanded depending on the services required. For instance, in order to support a military justice team, a single attorney cannot represent both the prosecution and defense. Therefore, in this case, the task organization would expand to two attorneys with each having his or her own clerk.

Field Exchange

Exchange services are provided to Marines both in garrison and while deployed through MCCS. In a deployed environment, the amount of services and merchandise offered is more limited, but it will provide support for Marines and their needs in theater. The primary unit of employment for field services is the tactical field exchange, which provides all the goods and services for Marines in theater. One MCCS officer and three MCCS Marines staff the field exchange. It requires one quadcon for all of its stock. A tactical field exchange has the capacity of supporting 1,000 customers per day. This is based on nine hours of service and three hours of stocking. For every additional 1,000 Marines who need to be supported, two more MCCS Marines are required. A mobile field exchange is employed as part of the WES team discussed earlier in this chapter.

Disbursing

The primary tasks supported by the disbursing section in a deployed environment are claims, currency exchange, personnel pay, and check cashing. Similar to the other functions discussed, disbursing is normally conducted by task-organized detachments that are contingent on the mission, size of unit supported, and duration of mission. Usually a disbursing section will employ one of two units of employment. The first is a procurement team. This team consists of a contingency contracting specialist, a supply administrative clerk, and a finance technician. This team is responsible for all tasks that support the procurement process. The other unit of employment is a field disbursing office. This office comprises two financial management officers and three financial technicians. In a deployed environment, a team of this size has the capacity to support two to three locations and provides mobile disbursing support, such as personnel pay.

Food services

Food services is a command support service function as opposed to a CSSE service function like the others previously discussed. However, because food services is a key logistical function, it is included as a part of this section to demonstrate logistics capacity. Field messing is the task of providing nutritional meals to military personnel. Field service Marines primarily perform this task. The principle equipment associated with messing is the expeditionary field kitchen (EFK), enhanced tray ration heating system (ETRHS), and the tray ration heating system (TRHS). The planning factors for number of personnel served per piece of equipment is 750 for the EFK, 250 for the ETRHS, and 250 for the TRHS. The number of field services Marines required for employment is based on the type of equipment utilized. Six to eight Marines are required to operate the EFK, while only two are required to run the ETRHS and TRHS.

Summary

The LUEs are a useful tool to assist logisticians in articulating capacity to the supported commander. Each LUE demonstrates a task the MLG is expected to perform and clearly articulates the personnel and equipment required to accomplish that task. This building-block approach is easily able to scale up to demonstrate unit capacity. However, there are a few recommendations for how the LUEs could be better refined.

The metrics used to define capacity need to be useful to the supported commander. Most metrics associated with an LUE reflect a unit produced over time (e.g., gallons per hour). However, while easy to compute for some LUEs, it does not always reveal anything useful about what logistics capacity is being provided. For instance, the amount of kW power produced is easy to calculate based on the type of generators employed by a power generation team. However, using a metric, such as size of structure powered, might provide a metric the supported commander better understands. Although these metrics are more subjective, they might be more useful.

While the LUEs identify the personnel required to conduct a task, they do not account for such factors as appropriate grade and required training. These were determined to be measures of capability, where the LUE measures capacity. To determine capability, further assessment of personnel qualifications should be conducted. Similarly, it is difficult within the current LUE structure to easily identify substitute equipment that could perform the task similarly, but perhaps not to its optimal output. While the LUE captures all equipment, this variation in output is not always clear. In addition, the LUEs in their present form do not identify personnel and equipment in high demand across multiple LUEs (e.g., MHE operators). However, these high-demand assets become more noticeable when the T/Os of units are assessed to determine how many LUEs can be supported. This assessment leads to the discovery of which pieces of equipment or personnel are limiting factors in developing multiple LUEs.

As mentioned earlier, determining capacity for maintenance and supply proved to be the most difficult because of the requirement for both organic DS LUEs and broader GS LUEs. While a redline approach proved useful, further validation is required.

Finally, LUEs are only as strong as the data that support them. In order to determine capacity, the Marine Corps needs to do a better job at tracking the raw data of logistics performance. For example, when speaking with supply SMEs, we were able to see the past few years' worth of requisition data for certain units, as well as previous manning documents. These data helped determine the appropriate size for an organic supply detachment and estimate the average capacity by analyzing historical data. That being said, even these data were limited. A systematic approach to data collection would be beneficial in supporting further refinement of the LUEs.

CHAPTER FIVE

Logistics Units of Employment Framework

This chapter discusses the use of the LUEs as a framework for estimating logistics requirements and capacity. A *framework* is taken here to mean a basic structure underlying a system or concept. It is intended to support or guide the extension of the basic structure into useful constructs. In this respect, the LUEs themselves represent a basic, extendable framework.

Two potential applications of the LUEs as a framework are illustrated. The first is using LUEs to approximate the capacity of a Marine Corps logistics organization. The second is using LUEs to understand the logistics capacity required for MEU CLB support to HADR.

Measuring the Capacity of CLB-11

We chose CLB-11 from 1st MLG to demonstrate the use of the LUEs as capacity metrics. The process involved examining the T/O and T/E for CLB-11 and assessing the number of LUEs that it could support. Appendix A lists the entire set of LUEs, and Appendix B describes the T/O and T/E spreadsheets for CLB-11. The files containing the LUEs and the T/O and T/E for CLB-11 are downloadable from the RAND website. The spreadsheet used to calculate the number of LUEs within CLB-11 (Logistics Units of Employment CLB Verification Version 4.xlsx) was also provided electronically to the sponsor.

The following tables show the resulting number and type of each LUE supportable by CLB-11's structure, arranged by function. The LUE number is an identifier created by the RAND team to facilitate the tracking of LUEs. Table 5.1 identifies which of the 12 possible transportation LUEs are present in CLB-11, and how many of each LUE CLB-11 can support.

TRANSPO10, the A/DACG LUE, serves as an example of how the LUEs can be used to determine the logistics capacity of CLB-11. As shown in the LUE spreadsheet, one TRANSPO10 LUE requires

- nine landing support Marines (MOS 0481)
- three engineer equipment operators (MOS 1345)
- one tractor, rubber tired, articulated steering machine (TRAM)
- one forklift with capacity to lift 5,000 pounds
- one extendable boom forklift (EBFL).

Table 5.1
Transportation LUEs Supported by CLB-11

LUE Number	Number of LUEs Supported	LUE Name	Total Capacity (Across CLB-11)
TRANSPO1		MTT–Convoy	
TRANSPO2	37	MTT–Passenger	506 passengers
TRANSPO3	30	MTT–Cargo	441.5 short tons
TRANSPO4	2	MTT–Water Distribution	3,600 gallons
TRANSPO5	~30	MTT–Fuel Distribution	27,000 gallons
TRANSPO6	3	HSTs	3 points/15,000–40,000 pounds per lift
TRANSPO7	1	POG	
TRANSPO8	1	BOG	1 beach
TRANSPO9	1	RHOG	
TRANSPO10	1	A/DACG	
TRANSPO12	7	Materiel Handling	5,000–10,000 lift

CLB-11's T/O shows that it has 26 LS Marines and seven engineer equipment operators; and the unit's T/E shows that it has three TRAMs, three forklifts with capacity to lift 5,000 pounds, and one EBFL. CLB-11 has the personnel, trams, and forklifts to support more than one TRANSPO10 LUE. However, it only has one EBFL and therefore can support only one LUE. The single EBFL is what limits CLB-11's capacity for A/DACG in this case.

TRANSPO11, the air delivery LUE, is not supported by CLB-11. This is because one air delivery LUE requires four airborne and aerial delivery specialists (MOS 0451) and one hospital corpsman (MOS 8404). However, there are no airborne and aerial delivery specialists within CLB-11. Therefore, the unit lacks airborne and aerial delivery capability. Having 13 hospital corpsmen does not help in this case.

Table 5.2 shows the general engineering LUEs supported by CLB-11.

Table 5.2
General Engineering LUEs Supported by CLB-11

LUE Number	Number of LUEs Supported	LUE Name	Total Capacity (Across CLB-11)
ENG3	1	Horizontal Construction Team	
ENG4	1	Vertical Construction Team	
ENG5	2	Electrical Power Distribution Teams	616 kW power
ENG6	3	Air Conditioning Teams	312,000 BTUs
ENG7	2	Water Purification Teams	3,000 gallons per hour
ENG8	2	Water Distribution Points	51,600 gallons
ENG9	2	Shower Teams	24 shower points for 300 Marines per hour
ENG12	2	Forward Vehicle Refueling Point	24,000 gallons of fuel
ENG13	4	EOD Teams	

We counted the general engineering LUEs supportable by CLB-11 using the same method of comparing the personnel and equipment requirements for each LUE against the T/O and T/E. Here, CLB-11 does not have sufficient personnel of the right specialties to support either the bridging team for a tactical bridge (ENG1) or the bridging team for a floating bridge (ENG2). The necessary equipment also did not appear to be present in the T/E.

Table 5.3 identifies the number of supply LUEs supported by CLB-11. The personnel in the T/O were only sufficient to support two LUEs.

In the case of supply administrative clerks (MOS 3043), there were 10 available for all of CLB-11, but multiple LUEs called for them. Each supply detachment (SUPP1) required four supply administrative clerks, meaning that the unit could support a total of two supply detachment LUEs. Other LUEs, such as general account (SUPP4), deployment support (SUPP5), initial issue provisioning (SUPP6), and customer service (SUPP11), all required more than 10 supply administrative clerks to support even a single LUE. Therefore, CLB's capacity for all these LUEs was zero. However, this raises the question of how to account for different combinations of LUEs that might be supportable with the same personnel and equipment available to a unit. This is especially the case for units larger than this example, where higher numbers of personnel and equipment might offer greater choices in the way they might be distributed across LUEs.

Table 5.4 lists the maintenance LUEs supported by CLB-11. The LUEs listed are predominantly at the organizational LOM, rather than at higher echelons of maintenance. This is consistent with the overall

Table 5.3
Supply LUEs Supported by CLB-11

LUE Number	Number of LUEs Supported	LUE Name	Total Capacity (Across CLB-11)
SUPP1	2	Supply Detachments	250 requisitions per month
SUPP2	5	Field ASP	5 sites

Table 5.4
Maintenance LUEs Supported by CLB-11

LUE Number	Number of LUEs Supported	LUE Name	Total Capacity (Across CLB-11)
MAINT1	1	Vehicle Recovery–AAV	
MAINT2	1	Vehicle Recovery–Tank	
MAINT3	2	Vehicle Recovery–Wheeled	
MAINT4	7	Maintenance Contact Team–Ordnance	
MAINT5	8	Maintenance Contact Team–Communications/ Electric	
MAINT6	2	Maintenance Contact Team–MT	
MAINT7	4	Maintenance Contact Team–Engineer	
MAINT8	2	Maintenance Support Team	

organizing principle for the MEU CLB, which has to rely on maintenance capability outside of it for more-advanced repair and specialized maintenance support.

CLB-11's T/O and T/E support the health services LUEs listed in Table 5.5. These health services LUEs are those for which CLB-11 has the appropriate personnel in its T/O. However, for ERCS (HS2) and field dental (HS10), the corresponding authorized allowance lists are not listed in the T/E. Therefore, these two are noted with a question mark. Additionally, there is one critical care nurse (MOS 2900) in CLB-11's T/O, but both the ERCS (HS2) and STP (HS4) LUEs call for one.

Table 5.6 presents the final table of LUEs supported by CLB-11, for services. The two services LUEs supported by the unit are the mobile-unit post office and field food services.

There are some limitations to applying LUEs to an official T/O and T/E to estimate logistics capabilities and capacities. The first is that

Table 5.5
Health Services LUEs Supported by CLB-11

LUE Number	Number of LUEs Supported	LUE Name	Total Capacity (Across CLB-11)
HS2	1?	ERCS? (Missing AMAL)	2 critically injured but stabilized for 2 hours
HS3	1	Ambulance Detachment	6 litter for 12 ambulatory patients
HS4	1	STP	50 trauma cases
HS9	1	Preventive Medicine Section	1 site
HS10	1?	Field Dental? (Missing ADAL)	200 patients for 30 days

Table 5.6
Services LUEs Supported by CLB-11

LUE Number	NUmber of LUEs Supported	LUE Name	Total Capacity (Across CLB-11)
SERV1	1	Mobile-Unit Post Office	500 Marines served
SERV12	2	Field Food Services	500-plus Marines fed

official T/Os and T/Es often do not reflect actual manning or the status of equipment. Actual manning often departs from the T/O; in many cases, coming in below full strength. LUEs should be applied to actual manning and available equipment to better estimate the true logistics capacity. Additionally, units will sometimes temporarily borrow equipment and personnel to achieve certain capabilities, and this would not be evident on their T/O or T/E.

Yet even with these limitations, the LUEs offer a potential way to standardize logistics capacity measures across different logistics organizations. Comparing the LUEs that different units should be able to support provides a first-order approximation of logistics capacities across organizational structures. It is also possible to extend this to

1st MLG's entire structure: Adding up all the LUEs that its total structure can support will produce an approximate total capacity in each logistics functional area. It is then possible to see how many LUEs are dedicated to deployed forces, overall MEF support, or are in residual capacity at a given point in time.

Measuring Logistics Capacity for MEU CLB Support to HADR

We can also use the LUEs to estimate the likelihood that a given logistics organizational structure is sufficient to support a mission. We do this by comparing the types of LUEs needed for a mission with the unit's available LUEs.

Table 5.7 offers a summary of the unique LUEs identified for the HADR mission. General LCE tasks and C2-related tasks for logistics did not have an LUE tied to them. These tasks still need to be accomplished but are not covered by the LUEs in Table 5.7. Chapter One discusses the method used to associate the MCTs, logistics-related T&R standards for HADR tasks, and the assignment of appropriate LUEs. Table 5.7 arranges the identified LUEs by functional area of logistics. Note that no LUEs were identified using this method for the services functional area of logistics. Appendix C goes into greater detail by associating LUEs to individual logistics tasks for HADR. Appendix C lists the T&R standard, the tasks, the LUE identifier, and descriptions of the LUEs for each task.

Table 5.7 indicates that a MEU CLB preparing for HADR may generally be expected to need such LUEs as an HST (TRANSPO6), horizontal construction team (ENG3), ASP (SUPP2), and an ambulance section (HS3). However, Table 5.7 does not contain information about the *quantity* of each LUE that may be required for a mission. For example, MEU CLB support to HADR requires a horizontal construction team. Appendix C gives additional information that at least six different tasks require horizontal construction. However, it is beyond the scope of this report to say how *many* are necessary for a mission.

Table 5.7
LUEs Associated with MEU CLB Support to HADR

Functional Logistics Area	LUE Identifiers	LUE Descriptions
Transportation	TRANSPO2	MTT–People
	TRANSPO3	MTT–Water Distribution
	TRANSPO4	MTT–Fuel Distribution
	TRANSPO5	MTT–Bulk Cargo
	TRANSPO6	HST
	TRANSPO7	POG
	TRANSPO8	BOG
	TRANSPO12	MHE
General Engineering	ENG3	Construction Team–Horizontal
	ENG4	Construction Team–Vertical
	ENG5	Electric Power Distribution
	ENG6	Air Conditioning
	ENG12	Forward Vehicle Refueling Point
Supply	SUPP2	ASP
	SUPP7	Fiscal/Procurement
	SUPP8	Property Accounting
	SUPP13	Supply Distribution
Maintenance	MAINT1	Vehicle Recovery Team–AAV
	MAINT2	Vehicle Recovery Team–Tank
	MAINT3	Vehicle Recovery Team–Wheeled
	MAINT4	Maintenance Contract Team–EOD
	MAINT9	MT Maintenance
	MAINT10	Engineer Maintenance
	MAINT11	Ordnance Maintenance
	MAINT12	Communications and Electric
	MAINT13	Maintenance; Repairable Maintenance
Health Services	HS3	Ambulance Section
	HS5	FRSS
	HS10	Field Dental Clinic
	HS11	BAS

The number would depend on the demands and magnitude of the particular situation and how those teams would be employed.

Because of the task-organized nature of the MAGTF, it may never make sense to specify standardized packages of personnel and equipment for logistics support to actual operations. For the purposes of providing additional information for readiness or force management, however, it may be useful to estimate the number of missions that a given structure might be able to support. In other words, taking Table 5.7 and adding an estimated quantity of each LUE would then allow an analyst to estimate the number of HADR missions a CLB or MLG might be able to support by examining the T/O and T/E. (It would also be possible to make assumptions about high, low, and medium-size operations.) It is important to reiterate that such an extension of Table 5.7 would not be to dictate LCE task organization or to remove flexibility from logistics planning, but to allow better understanding of issues such as residual capacity of the MLG.

There are other limitations with using Table 5.7 to inform logistics capacity requirements. The trip to 1st MLG in January 2016 elicited numerous logistics-related tasks from SMEs for the HADR mission presented by the RAND study team. Although there was insufficient time within the scope of the current project to review all the generated tasks, the number and range of tasks suggests that using the MCTs and T&R standards might underestimate the true variety of logistics tasks that the MEU CLB can be called upon to support for HADR and other missions by extension. Examples of potential logistics tasks identified during the brainstorming session with 1st MLG, but not necessarily in the MCTs, include providing mass casualty assistance, ordering different classes of supplies, establishing unclassified and classified computer networks, and providing laundry service, to name but a few. Consequently, the LUEs listed in Table 5.7 and Appendix C may underestimate the full range of LUEs actually involved in a HADR mission, and merely constitute the starting point for understanding LUEs that are likely needed.

Measuring Logistics Capacity for MEU CLB Support to NEO

The T&R manual for MEU support to NEO specifies one task for the LCE: "provide evacuation control center operations."[1] No LUE was an obvious fit for this task. Therefore, there is not an equivalent list

---

1 Headquarters, U.S. Marine Corps, 2012a, p. 2-12.

of mission-related LUEs as illustrated for the HADR in Table 5.7 and Table C.1. While there are most likely additional logistics tasks required to conduct NEO, they have not yet been codified and included in the T&R manual. The lack of specific tasks in the T&R manual made it difficult to assess the required LUEs in a similar manner to the HADR mission.

CHAPTER SIX

Conclusions and Recommendations

Conclusions

Based on the results of this report, we found that it is feasible to create a framework for understanding logistics capacity for the Marine Corps. The LUEs represent the basic building blocks of logistics capacity for the Marine Corps—the smallest set of personnel and equipment needed to accomplish essential tasks. The LUEs can be applied across the functional areas of Marine Corps logistics, and can be used as a basic, extendable framework to quantify logistics capacity across the MLG. Because a given logistic unit's T/O and T/E can be expressed as LUEs, the LUEs provide a means to gauge and compare the capacity of different units.

This report also demonstrates the hierarchy of frameworks that can be applied for different purposes and at different organizational levels. In addition, it is also possible to identify the LUEs associated with different missions as an extension of the basic LUE framework. Identifying the LUEs needed for a mission with the LUEs available from within an organization provides information on how many missions of what type the organization might support. The extension of the basic LUE framework to HADR missions illustrates this point. The RAND team was unable to apply the framework to the NEO mission because of limited information.

Recommendations

This study represents a promising start on developing a capacity assessment framework for MLGs or any LCE across the Marine Corps. However, the framework was tested on a single unit (CLB-11) and two missions (HADR and NEO). Much more remains to be accomplished before the LUEs listed in Appendix A can be validated. In the future, we suggest that additional work be accomplished in these areas:

- **Measure capability:** The current version of the LUEs does not take grade or training requirements into consideration when accounting for personnel. They also have limited information about substitutes for the equipment listed. These factors were determined to be measures of capability and so were not included in the LUE. Continuing the assessment by adding these components would enable assessments of capability.
- **SME LUE input:** A significant limitation of the LUEs is the nature of the SME input used to create many of them. In cases where there was little written guidance on what personnel and equipment were necessary to accomplish a task, we relied heavily on SME judgment. However, the SMEs used in this study were limited to only a very narrow sample of Marine Corps logistics experts; wider sampling of experts and examining the actual employment of personnel and equipment is needed.
- **Expand the application of the LUE framework:** Apply the basic LUE framework to the entire T/O and T/E of 1st MLG or another MLG. A logical next step is to turn the LUEs into a logistics estimation tool or dashboard that would demonstrate capacity based on current unit resources, which the Marine Corps is currently developing. By gathering information on the number of LUEs needed to support the MEF outside of deployments, we would also better understand residual capacity.
- **Expand the mission set:** Identify the approximate number of LUEs needed for each CLB mission. This would provide valuable information in assessing readiness and force management

and would better assess the number of missions an MLG might potentially support.

- **Quantify capacity:** Develop a better method to quantify capacity in harder-to-measure areas, such as the supply and maintenance functional areas.
- **Logistics C2:** This is not addressed in this study. Potential limitations stemming from C2 issues are not currently accounted for in the LUEs, but it would be an important issue to examine in greater detail.
- **Self-sustainment:** Account for the logistics support that logistics units require to sustain themselves, even as they provide support to other Marine units and entities outside the Marine Corps. Because this is not currently captured, the true logistics capacity required for the Marine Corps might be systematically underestimated.

APPENDIX A

Logistics Units of Employment

This appendix lists the LUEs by functional area. There is one table per functional area, which lists the LUE identifier number, the LUE, subfunctional area, the associated MET, and the MCT. The full Excel spreadsheet of LUEs (Logistics Units of Employment_v4.xlsx) is downloadable from the RAND website. Additional information in the Excel spreadsheet for each LUE includes key MOS(s) and numbers of personnel, key equipment, and quantities of key equipment.

Table A.1
Transportation

LUE Number	LUE	Function and Subfunction of Tactical Logistics	MET	MCT Number
TRANSPO1	MTT	Transportation–Passenger/Freight	Conduct Transportation Operations	MCT 4.3
TRANSPO2	MTT–People	Transportation–Passenger	Conduct MT Operations	MCT 4.3.3
TRANSPO3	MTT–Water Distribution	Transportation–Freight	Conduct MT Operations	MCT 4.3.3
TRANSPO4	MTT–Fuel Distribution	Transportation–Freight	Conduct MT Operations	MCT 4.3.3
TRANSPO5	MTT–Bulk Cargo	Transportation–Freight	Conduct MT Operations	MCT 4.3.3

Table A.1—Continued

LUE Number	LUE	Function and Subfunction of Tactical Logistics	MET	MCT Number
TRANSPO6	HST	Transportation–LS	Conduct Port and Terminal Support; Conduct LS Operations	MCT 4.3.2; MCT 4.3.9
TRANSPO7	POG	Transportation–LS	Conduct Port and Terminal Support; Conduct LS Operations	MCT 4.3.2; MCT 4.3.9
TRANSPO8	BOG	Transportation–LS	Conduct Port and Terminal Support; Conduct Landing Support Operations	MCT 4.3.2; MCT 4.3.9
TRANSPO9	RHOG	Transportation–LS	Conduct Port and Terminal Support; Conduct LS Operations	MCT 4.3.2; MCT 4.3.9
TRANSPO10	A/DACG	Transportation–LS	Conduct Port and Terminal Support; Conduct LS Operations	MCT 4.3.2; MCT 4.3.9
TRANSPO11	Air Delivery	Transportation–Air Delivery	Conduct Air Delivery	MCT 4.3.4
TRANSPO12	Materiel Handling	Transportation–Materiel Handling	Conduct Materiel Handling Operations	MCT 4.3.6

Table A.2
General Engineering

LUE Number	LUE	Function and Subfunction of Tactical Logistics	MET	MCT Number
Bridging				
ENG1	Bridging Team–Tactical Bridge	Engineering–Mobility Operations	Conduct Mobility Operations	MCT 1.4.1
ENG2	Bridging Team–Floating Bridge	Engineering–Mobility Operations	Conduct Mobility Operations	MCT 1.4.1
General Engineering				
ENG3	Construction Team–Horizontal	Engineering–General Engineering	Conduct General Engineering Operations	MCT 4.4
ENG4	Construction Team–Vertical			
Utilities				
ENG5	Electrical Power Distribution	Engineering–Electrical Supply	Conduct Tactical Electrical Supply	MCT 4.4.4
ENG6	Air Conditioning	Engineering–Electrical Supply	Conduct Tactical Electrical Supply	MCT 4.4.4
Bulk Liquid				
ENG7	Water Purification	Engineering–Bulk Liquid	Conduct Bulk Liquid Operations	MCT 4.4.3
ENG8	Water Distribution Point	Engineering–Bulk Liquid	Conduct Limited Bulk Liquid Operations	MCT 4.4.3.1
ENG9	Shower Team	Engineering–Bulk Liquid	Conduct Bulk Liquid Operations	MCT 4.4.3
ENG10	Laundry Services	Engineering–Bulk Liquid	Conduct Bulk Liquid Operations	
ENG11	Expeditionary Fuel Support–Bulk Fuel Platoon	Engineering–Bulk Liquid	Conduct Bulk Liquid Operations	MCT 4.4.3

Table A.2—Continued

LUE Number	LUE	Function and Subfunction of Tactical Logistics	MET	MCT Number
ENG12	Forward Vehicle Refueling Point	Engineering–Bulk Liquid	Conduct Limited Bulk Liquid Operations	MCT 4.4.3.1
EOD				
ENG13	EOD Section	Engineering–EOD	Conduct EOD Operations	MCT 6.8

Table A.3
Supply

LUE Number	Logistics Unit of Employment	Function and Subfunction of Tactical Logistics	MET	MCT Number
Organic Supply				
SUPP1	Supply Detachment	Supply	Conduct Ground Supply Operations	MCT 4.1.2
SUPP2	Field ASP	Supply	Conduct Ground Supply Operations	MCT 4.1.2
SUPP3	Expeditors	Supply	Conduct Ground Supply Operations	MCT 4.1.2
Intermediate Supply				
SUPP4	General Accounting	Supply	Conduct Ground Supply Operations	MCT 4.1.2
SUPP5	Deployment Support	Supply	Conduct Ground Supply Operations	MCT 4.1.2
SUPP6	Initial Issue Provisioning	Supply	Conduct Ground Supply Operations	MCT 4.1.2
SUPP7	Fiscal and Procurement	Supply–Procurement	Conduct Procurement	MCT 4.1.2.2
SUPP8	Property Accounting	Supply	Conduct Ground Supply Operations	MCT 4.1.2
SUPP9	Warehousing and Storage	Supply–Storage	Provide Storage	MCT 4.1.2.3
SUPP10	Packing			
SUPP11	Customer Service			
SUPP12	Rations			
SUPP13	Supply Distribution	Supply	Conduct Ground Supply Operations	MCT 4.1.2
SUPP14	Medical Logistics			
SUPP15	Ammunition Company			
SUPP16	Deployed Ammunition Support Team	Supply	Conduct Ground Supply Operations	MCT 4.1.2

Table A.4
Maintenance

LUE Number	Logistics Unit of Employment	Function and Subfunction of Tactical Logistics	MET	MCT Number
Organic Level–Vehicle Recovery Teams				
MAINT1	Vehicle Recovery Team–AAV	Maintenance–Recovery	Conduct Recover and Evacuation Operations	MCT 4.2.2.8
MAINT2	Vehicle Recovery Team–Tank	Maintenance–Recovery	Conduct Recover and Evacuation Operations	MCT 4.2.2.8
MAINT3	Vehicle Recovery Team–Wheeled	Maintenance–Recovery	Conduct Recover and Evacuation Operations	MCT 4.2.2.8
Organic Level–Maintenance Contact Teams				
MAINT4	Maintenance Contact Team–Ordnance	Maintenance–Repair	Conduct Repair	MCT 4.2.2.4
MAINT5	Maintenance Contact Team–Engineer	Maintenance–Repair	Conduct Repair	MCT 4.2.2.4
MAINT6	Maintenance Contact Team–Communications/Electronics	Maintenance–Repair	Conduct Repair	MCT 4.2.2.4
MAINT7	Maintenance Contact Team–MT	Maintenance–Repair	Conduct Repair	MCT 4.2.2.4
MAINT8	Maintenance Support Team	Maintenance–Repair	Conduct Repair	MCT 4.2.2.4
Field-Level Maintenance				
MAINT9	MT Maintenance	Maintenance–Repair	Conduct Repair	MCT 4.2.2.4
MAINT10	Engineer Maintenance			

Table A.4—Continued

LUE Number	Logistics Unit of Employment	Function and Subfunction of Tactical Logistics	MET	MCT Number
MAINT11	Ordnance Maintenance			
MAINT12	Communications and Electronics Maintenance			
MAINT13	Repairable Management			

Table A.5
Health Services

LUE Number	Logistics Unit of Employment	Function and Subfunction of Tactical Logistics	MET	MCT Number
HS1	Holding Ward Section	Health Services–Casualty Holding	Conduct Temporary Casualty Holding	MCT 4.5.4
HS2	ERCS	Health Services–Casualty Evacuation	Conduct Casualty Evacuation	MCT 4.5.5
HS3	Ambulance Section	Health Services–Casualty Evacuation	Conduct Casualty Evacuation	MCT 4.5.5
HS4	STP	Health Services	Provide Health Services	MCT 4.5
HS5	FRSS	Health Services	Provide Health Services	MCT 4.5
HS6	Radiology Section	Health Services	Provide Health Services	MCT 4.5
HS7	Lab	Health Services	Provide Health Services	MCT 4.5
HS8	Medical Regulating Team	Health Services–Medical Regulating	Conduct Medical Regulating	MCT 4.5.8
HS9	Preventive Medicine	Health Services	Provide Health Services	MCT 4.5
HS10	Field Dental Clinic	Health Services–Dental	Provide Emergency Dental Services	MCT 4.5.7.3
HS11	BAS	Health Services	Provide Health Services	MCT 4.5

Table A.6
Services

LUE Number	Logistics Unit of Employment	Function and Subfunction of Tactical Logistics	MET	MCT Number
SERV1	Mobile-Unit Post Office	Support Services–Postal	Provide LCE Support Services	MCT 4.6.1
SERV2	Legal Services Support Team	Support Services–Legal	Provide LCE Support Services	MCT 4.6.1
SERV3	Legal Services Support Team (Military Justice)	Support Services–Legal	Provide LCE Support Services	MCT 4.6.1
SERV4	Legal Services Support Team (Command Advice)	Support Services–Legal	Provide LCE Support Services	MCT 4.6.1
SERV5	Legal Services Support Team (Legal Assistance)	Support Services–Legal	Provide LCE Support Services	MCT 4.6.1
SERV6	Legal Services Support Team (Operational Law)	Support Services–Legal	Provide LCE Support Services	MCT 4.6.1
SERV7	Legal Services Support Team (Fiscal Law)	Support Services–Legal	Provide LCE Support Services	MCT 4.6.1
SERV8	Legal Services Support Team (Detention Operations)	Support Services–Legal	Provide LCE Support Services	MCT 4.6.1
SERV9	Tactical Field Exchange	Support Services–Field Exchange	Provide LCE Support Services	MCT 4.6.1
SERV10	Procurement Team	Support Services–Disbursing	Provide LCE Support Services	MCT 4.6.1
SERV11	Expeditionary Disbursing Office	Support Services–Disbursing	Provide LCE Support Services	MCT 4.6.1
SERV12	Field Food Services	Support Services–Food Services	Provide LCE Support Services	MCT 4.6.1
SERV13	WES Team			

APPENDIX B

CLB-11 Tables of Organization and Equipment

The T/O and T/E for CLB-11 (11th MEU), 1st MLG are included as Excel spreadsheets and are downloadable from the RAND website:

- CLB-11 TO.xlsx contains the table of organization. It includes such fields as the billet identifier code, pay grade, billet description, chargeable category, and personnel count.
- CLB-11 TE.xlsx contains the list of major equipment for CLB-11. Fields in this spreadsheet include a description of the equipment and the quantity.

APPENDIX C

LUEs in Support of Humanitarian Assistance Disaster Relief

Table C.1 describes associated LUEs with the MCTs and logistics-related T&R standards needed for MEU CLB support to HADR. It maps the appropriate LUE and LUE identifier in Appendix A with individual logistics-related tasks. Overall MEU, C2, and a few other related tasks are also included for activities that might need logistics support but whose appropriate LUE was not immediately obvious.

Table C.1
LUEs Associated with T&R Standards for MEU CLB Support to HADR

Logistics-Related T&R Standard	Task	LUEs	Description
MEU-LCE-7001	Provide Combat Service Support		MEU
MEU-LCE-7003	Support Civil-Military Operations		MEU
MEU-LCE-7004	Support Disaster Relief Operations		MEU
MEU-LCE-7005	Support Noncombatant Evacuation Operations		MEU
MEU-LCE-4001	Provide Military Police Support		MEU
C2OP-OPS-7001	Communicate with Commander Throughout the Orders Process		C2

Table C.1—Continued

Logistics-Related T&R Standard	Task	LUEs	Description
C2OP-OPS-7003	Employ Command and Control Systems		C2
C2OP-OPS-7004	Execute Command and control of an Operations		C2
C2OP-PLAN-7007	Establish C2 Systems Integration Plan		C2
ENGR-CMOB-5701	Create Obstacles Group	ENG3	Construction Team–Horizontal
ENGR-MOBL-5401	Construct Expedient Helicopter Landing Zone	ENG3	Construction Team–Horizontal
ENGR-MOBL-5501	Construct Tactical Landing Zones	ENG3	Construction Team–Horizontal
ENGR-MOBL-5709	Construct Combat Roads	ENG3	Construction Team–Horizontal
ENGR-SURV-5401	Construct Survivability Positions	ENG4	Construction Team–Vertical
ENGR-SURV-5402	Harden Existing Structures	ENG4	Construction Team–Vertical
ENGR-SURV-5403	Construct Field Fortifications	ENG4	Construction Team–Vertical
ENGR-XENG-5401	Provide Engineer Equipment Support	TRANSPO12	Materiel Handling
ENGR-XENG-5402	Prepare Site For Construction	ENG3	Construction Team–Horizontal
ENGR-XENG-5501	Conduct Horizontal Construction	ENG3	Construction Team–Horizontal
ENGR-XENG-5701	Conduct Vertical Construction	ENG4	Construction Team–Vertical
FUEL-XENG-5502	Maintain Bulk Fuel Petroleum Distribution Site	ENG12	Forward Vehicle Refueling Point

Table C.1—Continued

Logistics-Related T&R Standard	Task	LUEs	Description
UTIL-XENG-5401	Provide Utilities Support	ENG5 ENG6	Electric Power Distribution Air Conditioning
FMS-HSS-5003	Provide Level I HSS	HS3 HS5 HS11	Ambulance Section FRSS BAS
FMS-HSS-4003	Provide Level I HSS	HS3 HS5 HS11	Ambulance Section FRSS BAS
FMS-FP-4012	Perform Dental Care	HS10	Field Dental Clinic
FMS-FP-4014	Perform Immunization	HS11	BAS
INF-C2-7XXX	Conduct COC Operations		Infantry
LOG-OPS-6004	Conduct Convoy Operations	TRANSPO2 TRANSPO3 TRANSPO4 TRANSPO5 MAINT3	MTT–People MTT–Water Transportation MTT–Fuel Distribution MTT–Bulk Cargo Vehicle Recovery Team–Wheeled
LOG-OPS-5004	Conduct Beach Operations	TRANSPO8	BOG
LOG-OPS-5005	Conduct Landing Force Support Party Operations	TRANSPO6 TRANSPO7 TRANSPO8	HST POG BOG
MCMT-MAIN-3002	Maintain Motor Transport Equipment	MAINT7 MAINT9	Maintenance Contact Team–MT MT Maintenance
MCMT-MAIN-3003	Perform Maintenance on Recovered Equipment	MAINT9 MAINT10 MAINT11 MAINT12 MAINT13	MT Maintenance Engineer Maintenance Ordnance Maintenance Communications and Electric Maintenance Repairable Maintenance
MCMT-MAIN-3005	Conduct Recovery Operations	MAINT3	Vehicle Recovery Team–Wheeled

Table C.1—Continued

Logistics-Related T&R Standard	Task	LUEs	Description
MCMT-OPER-3004	Conduct Convoy Operations	TRANSPO2 TRANSPO3 TRANSPO4 TRANSPO5 MAINT3	MTT–People MTT–Water Transportation MTT–Fuel Distribution MTT–Bulk Cargo Vehicle Recovery Team–Wheeled
MCMT-OPER-3006	Conduct Movement Control		
MCMT-OPER-3008	Conduct Refueling Operations	TRANSPO4	MTT–Fuel Distribution
ORDM-CSSO-3007	Employ Maintenance Contact Teams	MAINT4	Maintenance Contact Team–Ordnance
ORDM-CSSO-3008	Conduct Recovery of Ordnance Weapons Systems and Equipment	MAINT1 MAINT2	Vehicle Recovery Team–AAV Vehicle Recovery Team–Tank
SUPP-MISC-7003	Provide Munitions Supply	SUPP2	ASP
SUPP-MISC-7004	Provide Munitions Storage	SUPP2	ASP
SUPP-MISC-7005	Conduct Distribution Operations	SUPP13	Supply Distribution
SUPP-FISC-5003	Conduct Unit Fiscal Accounting for Commander	SUPP7	Fiscal and Procurement
SUPP-FISC-5004	Account for Property and Materiel	SUPP8	Property Accounting

NOTE: HSS = Health Service Support.

Abbreviations

AAV	amphibious assault vehicle
A/DACG	Arrival/Departure Airfield Control Group
ADAL	Authorized Dental Allowance List
AMAL	Authorized Medical Allowance List
ASP	ammunition supply point
BAS	battalion aid station
BOG	Beach Operations Group
BSSG	Brigade Service Support Group
BTU	British Thermal Unit
C2	command and control
CLB	Combat Logistics Battalion
CLP	Combat Logistics Patrol
CLR	Combat Logistics Regiment
COC	Combat Operations Center
COP	Combat Outpost
CSS	Combat Service Support
CSSE	Combat Service Support Element

CSSG	Combat Service Support Group
DS	direct support
EBFL	extendable boom forklift
ECC	Evacuation Control Center
EFK	expeditionary field kitchen
EOD	explosive ordnance disposal
ERCS	En Route Care Service
ESB	engineer support battalion
ETRHS	enhanced tray ration heating system
FOB	forward operating base
FRSS	Forward Resuscitative Surgical Section
FSSG	Force Service Support Group
GCE	Ground Combat Element
GS	general support
HADR	humanitarian assistance and disaster relief
HST	helicopter support team
I&L	Installations and Logistics
IDC	independent duty corpsman
kW	kilowatt
LCE	Logistics Combat Element
LOM	level of maintenance
LS	Landing Support
LUE	Logistics Unit of Employment

MAGTF	Marine Air Ground Task Force
MARFORCOM	Marine Forces Command
MCAS	Marine Corps Air Station
MCCS	Marine Corps Community Services
MCT	Marine Corps task
MCWP	Marine Corps Warfighting Publication
MEF	Marine Expeditionary Force
MEPDS	Mobile Electrical Power Distribution System
MET	mission-essential task
METL	mission-essential task list
MEU	Marine Expeditionary Unit
MGB	medium girder bridge
MHE	materiel-handling equipment
MLG	Marine Logistics Group
MOS	military occupational specialty
MP	military police
MT	motor transport
MTT	motor transport team
NEO	Noncombatant Evacuation Operation
OAD	Operations Analysis Directorate
OCS	Operational Contract Support
OEF	Operation Enduring Freedom
OIF	Operation Iraqi Freedom

PEI	principle end item
POG	Port Operations Group
RHOG	Rail Head Operations Group
SAC	Study Advisory Committee
SME	subject-matter expert
SPMAGTF	Special Purpose Marine Air Ground Task Force
StabOps	stability operations
STP	Shock Trauma Platoon
T/E	table of equipment
T/O	table of organization
T&R	training and readiness
TRAM	tractor, rubber tired, articulated steering machine
TRHS	tray ration heating system
TSB	transportation support battalion
TSG	transportation support group
WES	Warfighter Express Services

Glossary

The terms listed in Table G.1 are used throughout this report. The definitions listed are contextual, i.e., they apply to this document and research. Other definitions likely exist when used in other contexts.

Table G.1
Terms Used in this Report

Term	Definition
Logistics Unit of Employment (LUE)	The smallest U.S. Marine Corps force (combination of trained personnel and equipment) providing a specific capability within a logistics function.
Capability	Ability to employ capacity competently to generate output from a logistics system over time. Capability is measured in terms of the presence of the right personnel and right equipment needed to employ capacity.
Capacity	The measurement or estimate of the amount of output produced by a logistics system per unit of time.[a]
Framework	A basic structure underlying a system or concept. It is intended to serve as a support or guide for the building of something that expands the structure into something useful. The basic structure for this study is the LUE. It supports the construction of a unit capacity estimate.
Output	The work done or the amount produced by a logistics unit of employment over a period of time. Also referred to as the production rate.
Readiness	The ability of military forces to fight and meet the demands of assigned missions.[b]
Redline	Minimum personnel or equipment required for the mission.

Table G.1—Continued

[a] David S. Stoller, *Logistics Systems Capacity*, Santa Monica, Calif.: RAND Corporation, RM-4852-PR, 1966, p. 12.

[b] Joint Staff, U.S. Department of Defense, Joint Publication 1-02, *Department of Defense Dictionary of Military and Associated Terms*, Washington, D.C., November 8, 2010, p. 197.

Bibliography

1st Marine Logistics Group, homepage, October 2015. As of March 14, 2016: http://www.1stmlg.marines.mil

1st Marine Logistics Group representatives, in-person discussion with author Joslyn Hemler, January 8, 2016.

2nd Marine Logistics Group representative, telephone communication with authors Joslyn Hemler and Yuna Wong, September 17, 2015.

2nd Marine Logistics Group representative, telephone communication with authors Joslyn Hemler and Yuna Wong, September 29, 2015.

Benbow, Robert, and Patricia Neil, *MLG Reorganization: Focus and Analysis on the Headquarters Element*, Alexandria, Va.: Center for Naval Analyses, March 2007.

Headquarters, U.S. Marine Corps, *Logistics*, Marine Corps Doctrinal Publication 4, Washington, D.C., February 21, 1997.

———, *Maintenance Operations*, Marine Corps Warfighting Publication 4-11.4, Washington, D.C., April 24, 1998a.

———, *MAGTF Supply Operations*, Marine Corps Warfighting Publication 4-11.7, Washington, D.C., June 24, 1998b.

———, *Logistics Operations*, Marine Corps Warfighting Publication 4-1, Washington, D.C., April 15, 1999.

———, *Engineering Operations*, Marine Corps Warfighting Publication 3-17, Washington, D.C., February 14, 2000a.

———, *Tactical-Level Logistics*, Marine Corps Warfighting Publication 4-11, Washington, D.C., June 13, 2000b.

———, *Transportation Operations*, Marine Corps Warfighting Publication 4-11.3, Washington, D.C., September 5, 2001a.

———, *Services in an Expeditionary Environment*, Marine Corps Warfighting Publication 4-11.8, Washington, D.C., September 24, 2001b.

———, *Unit Embarkation Handbook*, Marine Corps Reference Publication 4-11.3G, Washington, D.C., December 10, 2004.

———, *Bulk Liquid Operations,* Marine Corps Warfighting Publication 4-11.6, Washington, D.C., June 19, 2005.

———, *Logistics Training and Readiness Manual*, Navy and Marine Corps 3500.27B, Washington, D.C., May 11, 2011.

———, *Marine Expeditionary Unit Training and Readiness Manual*, Navy and Marine Corps 3500.99, Washington, D.C., November 13, 2012a.

———, *Health Service Support Operations*, Marine Corps Warfighting Publication 4-11.1, Washington, D.C., December 10, 2012b.

———, *Marine Corps Field Feeding Program*, Marine Corps Reference Publication 4-11.8A, Washington, D.C., December 2, 2013.

———, *Ground Equipment Maintenance Program,* Marine Corps Order 4790.25, Washington, D.C., January 12, 2014a.

———, *Expeditionary Force 21*, Washington, D.C., March 2014b.

———, *Organization of Marine Corps Forces,* Marine Corps Reference Publication 5-12D, Washington, D.C., August 26, 2015.

Installations and Logistics representative, telephone communication with authors Joslyn Hemler and Yuna Wong, September 29, 2015.

Joint Staff, U.S. Department of Defense, Joint Publication 1-02, *Department of Defense Dictionary of Military and Associated Terms*, Washington, D.C., November 8, 2010.

———, Joint Publication 4-10, *Operational Contract Support*, Washington, D.C., July 16, 2014a.

———, Joint Publication 3-02, *Amphibious Operations,* Washington, D.C., July 18, 2014b.

Marin, Isabel, Christine Hannigan, Megan Misencik, and Clinton Jones, "Functionally Aligned Battalions." *Marine Corps Gazette*, Vol. 98, Issue 10, October 2014, pp. 24–27.

Marine Forces Command G-4 representative, telephone communication with authors Joslyn Hemler and Yuna Wong, September 21, 2015.

Perry, Walter L., Anthony Atler, Roald Euller, Angel R. Martinez, Todd Nichols, and Jonathan Welch, *Allocating Marine Expeditionary Unit Equipment to Minimize Shortfalls*, 3rd ed., Santa Monica, Calif.: RAND Corporation, TL-167-OSD, 2015. As of July 22, 2016:
http://www.rand.org/pubs/tools/TL167.html

Stoller, David S., *Logistics Systems Capacity*, Santa Monica, Calif.: RAND Corporation, RM-4852-PR, 1966. As of July 23, 2016:
http://www.rand.org/pubs/research_memoranda/RM4852.html

Usher, Edward G., John Sweeney, Darell Moore, Frank Tapia, et al., "Brute Force Combat Service Support: 1st Force Service Support Group in Operation Iraqi Freedom," *Marine Corps Gazette*, Vol. 87, Issue 8, August 2003, pp. 34–35.

U.S. General Accounting Office, *Measuring Military Capability: Progress, Problems, and Future Direction,* February 1986.

U.S. Marine Corps, Capabilities Development Directorate, *Marine Corps Task List 2.0,* Quantico, Va., September 1, 2016.